The Mechanism of Life

by Stephane Leduc

Translated W Deane Butcher

TRANSLATOR'S PREFACE

Professor Leduc's work has excited a good deal of attention, and not a little opposition, on the Continent. As recently as 1907 the Academie des Sciences excluded from its Comptes Rendus the report of these experimental researches on diffusion and osmosis, because it touched too closely on the burning question of spontaneous generation.

As the author points out, Lamarck's early evolutionary hypothesis was killed by opposition and neglect, and had to be reborn in England before it obtained universal acceptance as the Darwinian Theory. Not unnaturally, therefore, he turns for an appreciation of his work to the free air and wide horizon of the English-speaking countries.

He has entitled his book "The Mechanism of Life," since however little we may know of the origin of life, we may yet hope to get a glimpse of the machinery, and perhaps even hear the whirr of the wheels in Nature's workshop. The subject is of entrancing interest to the biologist and the physician, quite apart from its bearing on the question of spontaneous generation. Whatever view may be entertained by the different schools of thought as to the nature and significance of life, all alike will welcome this new and important contribution to our knowledge of the mechanism by which Nature constructs the bewildering variety of her forms.

There is, I think, no more wonderful and illuminating spectacle than that of an osmotic growth,--a crude lump of brute inanimate matter germinating before our very eyes, putting forth bud and stem and root and branch and leaf and fruit, with no stimulus from germ or seed, without even {viii} the presence of organic matter. For these mineral growths are not mere crystallizations as many suppose; they increase by intussusception and not by accretion. They exhibit the phenomena of circulation and respiration, and a crude sort of reproduction by budding; they have a period of vigorous youthful growth, of old age, of death and of decay. They imitate the forms, the colour, the texture, and even the microscopical structure of organic growth so closely as to deceive the very elect. When we find, moreover, that the processes of nutrition are carried on in these osmotic productions just as in living beings, that an injury to an osmotic growth is repaired by the coagulation of its internal sap, and that it is able to perform periodic movements just as an animal or a plant, we are at a loss to define any line of separation between these mineral forms and those of organic life.

In the present volume the author has collected all the data necessary for a complete survey of the mechanism of life, which consists essentially of those phenomena which are exhibited at the contact of solutions of different degrees of concentration. Whatever may be the verdict as to the author's case for spontaneous generation, all will agree that the book is a most brilliant and stimulating study, founded on the personal investigation of a born experimenter.

The present volume is a translation of Dr. Leduc's French edition, but it is more than this, the work has been translated, revised and corrected, and in many places re-written, by the author's own hand. I am responsible only for the English form of the treatise, and can but regret that I have been able to reproduce so imperfectly the charm of the original.

W. DEANE BUTCHER.

EALING.

TABLE OF CONTENTS

INTRODUCTION

Life was formerly regarded as a phenomenon entirely separated from the other phenomena of Nature, and even up to the present time Science has proved wholly unable to give a definition of Life; evolution, nutrition, sensibility, growth, organization, none of these, not even the faculty of reproduction, is the exclusive appanage of life.

Living things are made of the same chemical elements as minerals; a living being is the arena of the same physical forces as those which affect the inorganic world.

Life is difficult to define because it differs from one living being to another; the life of a man is not that of a polyp or of a plant, and if we find it impossible to discover the line which separates life from the other phenomena of Nature, it is in fact because no such line of demarcation exists--the passage from animate to inanimate is gradual and insensible. The step between a stalagmite and a polyp is less than that between a polyp and a man, and even the trained biologist is often at a loss to determine whether a given borderland form is the result of life, or of the inanimate forces of the mineral world.

A living being is a transformer of matter and energy--both matter and energy being uncreateable and indestructible, i.e. invariable in quantity. A living being is only a current of matter and of energy, both of which change from moment to moment while passing through the organism.

That which constitutes a living being is its form; for a living thing is born, develops, and dies with the form and structure of its organism. This ephemeral nature of the living being, which perishes with the destruction of its form, is in {xiv} marked contrast to the perennial character of the matter and the energy which circulate within it.

The elementary phenomenon of life is the contact between an alimentary liquid and a cell. For the essential phenomenon of life is nutrition, and in order to be assimilated all the elements of an organism must be brought into a state of solution. Hence the study of life may be best begun by the study of those physico-chemical phenomena which result from the contact of two different liquids. Biology is thus but a branch of the physico-chemistry of liquids; it includes the study of electrolytic and colloidal solutions, and of the molecular forces brought into play by solution, osmosis, diffusion, cohesion, and crystallization.

In this volume I have endeavoured to give as much of the science of energetics as can be treated without the use of mathematical formul? the conception of entropy and Carnot's law of thermodynamics are also discussed.

The phenomena of catalysis and of diastatic fermentation have for the first time been brought under the general laws of energetics. This I have done by showing that catalysis is only one instance of the general law of the transformation of potential into kinetic energy, viz. by the intervention of a foreign exciting and stimulating energy which may be infinitely smaller than the energy it transforms. This conception brings life into line with other catalytic actions, and shows us a living being as a store of potential energy, to be set free by an external stimulus which may also excite sensation.

In a subsequent chapter I have dealt with the rise of Synthetic Biology, whose history and methods I have described. It is only of late that the progress of physico-chemical science has enabled us to enter into this field of research, the final one in the evolution of biological science.

The present work contains some of the earliest results of this synthetic biology. We shall see how it is possible by the mere diffusion of liquids to obtain forms which imitate with the greatest accuracy not only the ordinary cellular tissues, but the more complicated striated structures, such as muscle and mother-of-pearl. We shall also see how it is {xv} possible by simple liquid diffusion to reproduce in ordered and regular succession complicated movements like those observed in the karyokinesis of the living cell.

The essential character of the living being is its Form. This is the only characteristic which it retains during the whole of its existence, with which it is born, which causes its development, and disappears with its death. The task of synthetic biology is the recognition of those physico-chemical forces and conditions which can produce forms and structures analogous to those of living beings. This is the subject of the chapter on Morphogenesis.

The last chapter deals with the doctrine of Evolution. The chain of life is of necessity a continuous one, from the mineral at one end to the most complicated organism at the other. We cannot allow that it is broken at any point, or that there is a link missing between animate and inanimate nature. Hence the theory of evolution necessarily admits the physico-chemical nature of life and the fact of spontaneous generation. Only thus can the evolutionary theory become a rational one, a stimulating and fertile inspirer of research. We seek for the physico-chemical forces which produce forms and structures analogous to those of living beings, and phenomena analogous to those of life. We study the alterations in environment which modify these forms, and we seek in the past history of our planet for those natural phenomena which have brought these physico-chemical forces into play. In this way we may find the road which will, we hope, lead some day to the discovery of the origin and the evolution of life upon the earth.

* * * * *

{1}

THE MECHANISM OF LIFE

CHAPTER I

LIFE AND LIVING BEINGS

Primitive man distinguished but two kinds of bodies in nature, those which were motionless and those which were animated. Movement was for him the expression of life. The stream, the wind, the waves, all were alive, and each was endowed with all the attributes of life--will, sentiment, and passion. Ancient Greek mythology is but the poetic expression of this primitive conception.

In the evolution of the intelligence, as in that of the body, the development of the individual is but a repetition of the development of the race. Even now children attribute life to everything that moves. For them a little bird still lives in the inside of a watch, and produces the tick-tick of the wheels. In modern times, however, we have learnt that everything in nature moves, so that motion of itself cannot be considered as the characteristic of life.

Heraclitus aptly compares life to a flame. Aristotle says, "Life is nutrition, growth, and decay,--having for its cause a principle which has its end in itself, namely [Greek: entelecheia]." This principle is itself in need of definition, and Aristotle only substitutes one unknown epithet for another.

Bichat defined life as the ensemble of the functions which resist death. This is to define life in terms of death,--but death is but the end of life, and cannot be defined without first defining life. Claude Bernard rejects all definition of life as insufficient, and incompatible with experimental science. {2}

Some modern physiologists regard sensibility, others irritability, as the characteristic of life, and define life as the faculty of responding, by some sort of change, to an external stimulus. As in the case of movement, we have found by more attentive observation that this faculty also is universal in nature. There is no action without reaction; an elastic body repels the body that strikes it. Every object in nature dilates with heat, contracts with cold, and is modified by the light which it absorbs. Everything in nature responds to exterior action by a change, and hence this faculty cannot be the characteristic of life.

A distinguished professor of physiology was accustomed to teach that the disproportion between action and reaction was the characteristic of life. "Allow a gramme weight to fall on a nerve, and the muscle will raise a weight of ten grammes. This disproportion is the characteristic of life." But there is a much greater disproportion between action and reaction when the friction of a match blows up a powder factory, or the turning of a switch lights the lamps and animates the tramways and the motors of a great city. The disproportion between action and reaction is therefore no characteristic of life.

The essential characteristic of life is often said to be nutrition--the phenomenon by which a living organism absorbs matter from its environment, subjects it to chemical metamorphosis, assimilates it, and finally ejects the destructive products of metamorphosis into the surrounding medium. But this characteristic is also common to a great number of ordinary chemical reactions, so that we cannot call it peculiar to life. Consider, for instance, a fragment of calcium chloride immersed in a solution of sodium carbonate. It absorbs the carbonic ion, incorporates it into a molecule of calcium carbonate, and ejects the chlorine ion into the surrounding medium.

It may be argued that this is merely a chemical process, since the substance which determines the reaction is also modified, the chloride of calcium changing into carbonate of calcium. But every living thing is also changing its chemical {3} constitution during every moment of its existence,--it is this change which constitutes the process of senile involution. The substance of the child is other than that of the ovum, and the substance of the adult is not that of the child. Hence we cannot regard nutrition as the exclusive characteristic of life.

Other authorities regard growth and organization as the essentials of life. But crystals also grow. It was said that the growth of a crystal differed from that of a living thing, in that the former grew by the addition of material from without--the juxtaposition of bricks, as it were--while the latter grew by intussusception, an introduction of fresh material into the substance of the organism. A crystal, moreover, was homogeneous, while the tissues of a living being were differentiated--such differentiation constituting the organization. At the present time, however, we recognize the existence of a great variety of purely physical productions, the so-called "osmotic growths," which increase by a process of intussusception, and develop therefrom a marvellous complexity of organization and of form. Hence growth and organization cannot be considered as the essential characteristics of life.

Since, then, we are totally unable to define the exact boundary which separates life from the physical phenomena of nature, we may fairly conclude that no such separation exists. This is in conformity with the "law of continuity,"--the principle which asserts that all the phenomena of nature are continuous in time and space. Classes, divisions, and separations are all artificial, made not by nature but by man. All the forms and phenomena of nature are united by insensible transition; it is impossible to separate them, and in the distinction between living and non-living things we must content ourselves with relative definitions, which are far from being precise.

Life can only be defined as the sum of all phenomena exhibited by living beings, and its definition thus becomes a mere corollary to the definition of a living being.

The true definition of a living being is that it is a transformer of energy, receiving from its environment the energy {4} which it returns to that environment under another form. All living organisms are transformers of energy.

A living organism is also a transformer of matter. It absorbs matter from its environment, transforms it, and returns it to its environment in a different chemical condition. Living things are chemical transformers of matter.

Living beings are also transformers of form. They commence as a very simple form, which gradually develops and becomes more complicated.

The matter of which a living organism is constituted consists essentially of certain solutions of crystalloids and colloids. To this we may add an osmotic membrane to contain the liquids, and a solid skeleton to support and protect them. Finally, it would seem that a colloid of one of the albuminoid groups is a necessary constituent of every living being.

We may say, then, that a living being is a transformer of energy and of matter, containing certain albuminoid substances, with an evolutionary form, the constitution of which is essentially liquid.

A living being has but a limited duration. It is born, develops, becomes organized, declines and dies. Through all the metamorphoses of form, of substance, and of energy, informing the whole course of its existence, there is a certain co-ordination, a certain harmony, which is necessary for the conservation of the individual. This harmony we call Life. Discord is disease,-- the total cessation of the harmony is Death. When the form is profoundly altered and the substance changed, the transformation of energy no longer follows its regular course, the organism is dead.

After death the colloids which have constituted the form of the living thing pass from their liquid state as "sols" into their coagulated state as "gels." The metamorphoses of form, substance, and energy still continue, but no longer harmoniously for the conservation of the individual, but in dis-harmony for its dissolution. Finally, the form of the individual disappears, the substance and the energy of the living being is resolved and dispersed into other bodies and other phenomena. {5}

The results hitherto obtained from the study of life seem but inconsiderable when compared with the time and labour devoted to the question. Max Verworn exclaims, "Are we on a false track? Do we ask our questions of Nature amiss, or do we not read her answers aright?"

Each branch of science at its commencement employs only the simpler methods of observation. It is purely descriptive. The next step is to separate the different parts of the object studied--to dissect and to analyse. The science has now become analytical. The final stage is to reproduce the substances, the forms, and the phenomena which have been the subject of investigation. The science has at last become synthetical.

Up to the present time, biology has made use only of the first two methods, the descriptive and the analytical. The analytical method is at a grave disadvantage in all biological investigations, since it is impossible to separate and analyse the elementary phenomena of life. The function of an organ ceases when it is isolated from the organism of which it forms a part. This is the chief cause of our lack of progress in the analysis of life.

It is only recently that we have been able to apply the synthetic method to the study of the phenomena of life. Now that we know that a living organism is but the arena for the transformation of energy, we may hope to reproduce the elementary phenomena of life, by calling into play a similar transformation of energy in a suitable medium.

Organic chemistry has already obtained numerous victories in the same direction, and the rapid advance in the production of organic bodies by chemical synthesis may be considered the first-fruits of synthetic biology.

A phenomenon is determined by a number of circumstances which we call its causes, and of which it is the result. Every phenomenon, moreover, contributes to the production of other phenomena which are called its consequences. In order therefore to understand any phenomenon in its entirety, we must determine all its causes both qualitatively and quantitatively.

Phenomena succeed one another in time as consequences {6} one of another, and thus form an uninterrupted chain from the infinite of the past into the infinite of the future. A living being gathers from its entourage a supply of matter and of energy, which it transforms and returns. It is part and parcel of the medium in which it lives, which acts upon it, and upon which it acts. The living being and the medium in which it exists are mutually interdependent. This medium is in its turn dependent on its entourage,--and so on from medium to medium throughout the regions of infinite space.

One of the great laws of the universe is the law of continuity in time and space. We must not lose sight of this law when we attempt to follow the metamorphoses of matter, of energy and of form in living beings. Evolution is but the expression of this law of continuity, this succession of phenomena following one another like the links of a chain, without discontinuity through the vast extent of time and space.

The other great universal law, that of conservation, applies with equal force to living and to inanimate things. This law asserts the uncreateability and the indestructibility of matter and of energy. A given quantity of matter and of energy remains absolutely invariable through all the transformations through which it may pass.

We need not here discuss the question of the possible transformation of matter into ether, or of ether into ponderable matter. Such a transformation, if it exists, would have but little bearing on the phenomena of life. Moreover, it also will probably be found to conform to the law of conservation of energy.

In marked contrast to the permanence of matter and of energy is the ephemeral nature of form, as exhibited by living beings. Function, since it is but the resultant of form, is also ephemeral. All the faculties of life are bound up with its form,--a living being is born, exists, and dies with its form.

The phenomena of life may in certain cases slow down from their normal rapidity and intensity, as in hibernating {7} animals, or be entirely suspended, as in seeds. This state of suspension of life, of latent life as it were, reminds us of a machine that has been stopped, but which retains its form and substance unaltered, and may be started again whenever the obstacle to its progress is removed.

During the whole course of its life a living being is intimately dependent on its entourage. For example, the phenomena of life are circumscribed within very narrow limits of temperature. A living organism, consisting as it does essentially of liquid solutions, can only exist at temperatures at which such solutions remain liquid, i.e. between 0°C. and 100°C. Certain organisms, it is true, may be frozen, but their life remains in a state of suspension so long as their substance remains solid. Since the albuminoid substances which are a necessary component of the living organism become coagulated at 44°C., the manifestations of life diminish rapidly above this temperature. The intensity of life may be said to augment gradually as the temperature rises from 0° to 40° and then to diminish rapidly as the temperature rises above that point, becoming nearly extinct at 60°C.

Another condition indispensable to life is the presence of oxygen. Life, compared by Heraclitus to a flame, is a combustion, an oxydation, for which the presence of oxygen at a certain pressure is indispensable. There are, it is true, certain anaobic micro-organisms which apparently exist without oxygen, but these in reality obtain their oxygen from the medium in which they grow.

Life is also influenced by light, by mechanical pressure, by the chemical composition of its entourage, and by other conditions which we do not as yet understand. In each case the conditions which are favourable or noxious vary with the nature of the organism, some living in air, some in fresh water, and others in the sea.

Formerly it was supposed that the substance of a living being was essentially different from that of the mineral world, so much so that two distinct chemistries were in existence--organic chemistry, the study of substances derived from bodies which had once possessed life, and inorganic chemistry, dealing {8} with minerals, metalloids, and metals. We now know that a living organism is composed of exactly the same elements as those which constitute the mineral world. These are carbon, oxygen, hydrogen, nitrogen, phosphorus, calcium, iron, sulphur, chlorine, sodium, potassium, and one or two other elements in smaller quantity. It was formerly supposed that the organic combinations of these elements were found only in living organisms and could be fashioned only by vital forces. In more recent times, however, an ever increasing number of organic substances have been produced in the laboratory.

Organic bodies may be divided into four principal groups. (1) Carbohydrates, including the sugars and the starches, all of which may be considered as formed of carbon and water. (2) Fats, which may be considered chemically as the ethers of glycerine, combinations of one molecule of glycerine and three molecules of a fatty acid, with elimination of water. (3) Albuminoids, substances whose molecules are complex, containing nitrogen and sulphur in addition to carbon, oxygen, and hydrogen. The albuminoid of the cell nucleus also contains phosphorus, and the hemoglobin of the blood contains iron. (4) Minerals or inorganic elements, such as chloride of sodium, phosphate of calcium, and carbonic acid. This group also includes water, which is the most important constituent, since it forms more than a moiety of the substance of all living creatures.

Wuhler in 1828 accomplished the first synthesis of an organic substance, urea, one of the products of the decomposition of albumin. Since then a large number of organic substances have been prepared by the synthesis of their inorganic elements. The most recent advance in this direction is that of Emile Fischer, who has produced polypeptides having the same reactions as the peptones, by combining a number of molecules of the amides of the fatty acids.

In the further synthesis of organic compounds the problems we have before us are of the same order as those already solved. There is no essential difference between organic and inorganic chemistry; living organisms are formed of the {9} same elements as the mineral world, and the organic combinations of these elements may be realized in our laboratories, just as in the laboratory of the living organism.

Not only so, but a living being only borrows for a short time those mineral elements which, after having passed through the living organism, are returned once again to the mineral kingdom from which they came.

All matter has life in itself--or, at any rate, all matter susceptible of incorporation in a living cell. This life is potential while the element is in the mineral state, and actual while the element is passing through a living organism.

Mineral matter is changed into organic matter in its passage through a vegetable organism. The carbonic acid produced by combustion and respiration is absorbed by the chlorophyll of the leaves under the stimulus of light--the oxygen of the carbonic acid being returned to the air, while the carbon is utilized by the plant for the formation of sugar, starch, cellulose, and fats.

Thus plants are fed in great part by their leaves, taking an important part of their nourishment from the air, while by their roots they draw from the earth the water, the phosphates, the mineral salts, and the nitrates required for the formation of their albuminoid constituents. A vegetable is a laboratory in which is carried out the process of organic synthesis by which mineral materials are changed into organic matter. The first synthetic reaction is the formation of a molecule of formic aldehyde, CH_2O, by the combination of a molecule of water with an atom of carbon.

From this formic aldehyde, or formol, we may obtain all the various carbohydrates by simple polymerization, i.e. by the association of several molecules, with or without elimination of water. Thus two molecules of formol form one molecule of acetic acid, $2CH_2O = C_2H_4O_2$. Three molecules of formol form a molecule of lactic acid, $3CH_2O = C_3H_6O_3$. Six molecules of formol represent glucose and levulose, $6CH_2O = C_6H_{12}O_6$. Twelve molecules of formol minus one molecule of water form saccharose, lactose, cane sugar, and sugar of milk, $12CH_2O = C_{12}H_{22}O_{11} + H_2O$; n times six $\{10\}$ molecules of formol minus one molecule of water, $n(C_6H_{10}O_5)$, form starch and cellulose.

Animals derive their nourishment from vegetables either directly, or indirectly through the flesh of herbivorous animals. The mineral matter, rendered organic in its passage through a vegetable growth, is finally returned by the agency of animal organisms to the mineral world again, in the form of carbonic acid, water, urea, and nitrates. Thus vegetables may be regarded as synthetic agents, and animals and microbes as agents of decomposition. Here also the difference is only relative, for in certain cases vegetables produce carbonic acid, while some animal organisms effect synthetic combinations. Moreover, there are intermediary forms, such as fungi, which possessing no chlorophyll are nourished like animals by organic matter, and yet like vegetables are able to manufacture organic matter from mineral salts.

The work of combustion begun by the animal organism is finished by the action of micro-organisms, who complete the oxydation--the re-mineralization of the chemical substances drawn originally from the inorganic world by the agency of plant life.

To sum up. Vegetables obtain their nourishment from mineral substances, which they reduce, de-oxydize, and charge with solar energy. Animal organisms on the contrary oxydize, and micro-organisms complete the oxydation of these substances, returning them to the mineral world as water, carbonates, nitrates, and sulphates.

Thus matter circulates eternally from the mineral to the vegetable, from the vegetable to the animal world, and back again. The matter which forms our structure, which is to-day part and parcel of ourselves, has formed the structure of an infinite number of living beings, and will continue to pursue its endless reincarnation after our decease.

This endless cycle of life is also an endless cycle of energy. The combination of carbon with water carried out by the agency of chlorophyll can only take place with absorption of energy. This energy comes directly from the sun, the red and orange light radiations being absorbed by the chlorophyll. {11} The arrest of vegetation during the winter months is due not so much to the lowering of temperature as to the diminution of the radiant energy received from the sun. In the same way shade is harmful to vegetation, since the radiant energy required for growth is prevented from reaching the plant.

The energy radiated by the sun is accumulated and stored in the plant tissues. Later on, animals feed on the plants and utilize this energy, excreting the products of decomposition, i.e. the constituents of their food minus the energy contained in it. Thus the whole of the energy which animates living beings, the whole of the energy which constitutes life, comes from the sun. To the sun also we owe all artificial heat, the energy stored up in wood and coal. We are all of us children of the sun.

The radiant energy of the sun is transformed by plants into chemical energy. It is this chemical energy which feeds the vital activity of animals, who return it to the external world under the form of heat, mechanical work, and muscular contraction, light in the glow-worm, electricity in the electric eel.

There is a marked difference between the forms affected by organic and inorganic substances. The forms of the mineral world are those of crystals--geometrical forms, bounded by straight lines, planes, and regular angles. Living organisms, on the contrary, affect forms which are less regular--curved surfaces and rounded angles. The physical reason for this difference in form lies in a difference of consistency, crystals being solid, whereas living organisms are liquids or semi-liquids. The liquids of nature, streams and clouds and dewdrops, affect the same rounded forms as those of living organisms.

Living beings for the most part present a remarkable degree of symmetry. Some, like radiolarians and star-fish, have a stellate form. In plants the various organs often radiate from an axis, in such a manner that on turning the plant about this axis the various forms are superposed thrice, four, or more often five times in one complete revolution. It is remarkable how often this number five recurs in the {12} divisions and parts of a living organism. In other cases the similar parts are disposed symmetrically on either side of a median line or plane, giving a series of homologous parts which are not superposable.

The most important characteristic of a living being is its form. This is implicitly admitted by naturalists, who classify animals and plants in genera and species according to the differences and analogies of their form.

All living beings are composed of elementary organizations called cells. In its complete state, a cell consists of a membrane or envelope containing a mass of protoplasm, in the centre of which is a nucleus of differentiated protoplasm. This nucleus may in its turn contain a nucleolus. In some cases the cell is merely a protoplasmic mass without a visible envelope, so that a cell may be defined as essentially a mass of protoplasm provided with a nucleus.

A living organism may consist merely of a single cell, which is able alone to accomplish all the functions of life. Most living beings, however, consist of a collection of innumerable cells forming a cellular association or community. When a number of cells are thus united to constitute a single living being, the various functions of life are divided among different cellular groups. Certain cells become specialized for the accomplishment of a single function, and to each function corresponds a different form of cell. It is thus easy to recognize by their form the nerve cells, the muscle cells which perform the function of movement, and the glandular cells which perform the function of secretion. The cells of a living being are microscopic in size, and it is remarkable that they never attain to any considerable dimensions.

In order that life may be maintained in a living organism, it is necessary that a continual supply of aliment should be brought to it, and that certain other substances, the waste-products of combustion, should be eliminated. In order to be absorbed and assimilated, the alimentary substances must be presented to the living organism in a liquid or gaseous state. Thus the essential condition necessary for the {13} maintenance of life is the contact of a living cell with a current of liquid. The elementary physical phenomenon of life is the contact of two different liquids. This is the necessary condition which renders possible the chemical exchanges and the transformations of energy which constitute life. It is in the study of the phenomena of liquid contact and diffusion that we may best hope to pierce the secrets of life. The physics of vital action are the physics of the phenomena which occur in liquids, and the study of the physics of a liquid must be the preface and the basis of all inquiry into the nature and origin of life.

* * * * *

{14}

CHAPTER II

SOLUTIONS

We have seen that living beings are transformers of energy and of matter, evolutionary in form and liquid in consistency; that they are solutions of colloids and crystalloids separated by osmotic membranes to form microscopic cells, or consisting merely of a gelatinous mass of protoplasm, with a nucleus of slightly differentiated material. The elementary phenomenon of life is the contact of two different solutions. This is the initial physical phenomenon from which proceed all the other phenomena of life in accordance with the ordinary chemical and physical laws. Thus the basis of biological science is the study of solution and of the phenomena which occur between two different solutions, either in immediate contact or when separated by a membrane.

A solution is a homogeneous mixture of one or more solutes in a liquid solvent. Before solution the solute or dissolved substance may be solid, liquid, or gaseous.

Solutes, or substances capable of solution, may be divided into two classes-- substances which are capable of crystallization, or crystalloids; and those which are incapable of crystallization, the colloids. Crystalloids may be divided again into two classes, those whose solutions are ionizable and therefore conduct electricity, chiefly salts, acids, and bases; and those whose solutions are non-ionizable and are therefore non-conductors. These latter are for the most part crystallizable substances of organic origin, such as sugars, urea, etc.

Avogadro's law asserts that under similar conditions of temperature and pressure, equal volumes of various gases {15} contain an equal number of molecules. Under similar conditions, the molecular weights of different substances have therefore the same ratio as the weights of equal volumes of their vapours. Hence if we fix arbitrarily the molecular weight of any one substance, the molecular weight of all other substances is thereby determined. The molecular weight of hydrogen has been arbitrarily fixed as two, and hence the molecular weight of any substance will be double its gaseous density when compared with that of hydrogen.

Gramme-Molecule.--A gramme-molecule is the molecular weight of a body expressed in grammes. Occasionally for brevity a gramme-molecule is spoken of as a "molecule." Thus we may say that the molecular weight of oxygen is 16 grammes, meaning thereby that there are the same number of molecules in 16 grammes of oxygen as there are atoms in 1 gramme of hydrogen.

Concentration.--The concentration of a solution is the ratio between the quantity of the solute and the quantity of the solvent. The concentration of a solution is expressed in various ways. (a) The weight of solute dissolved in 100 grammes of the solvent. (b) The weight of solute present in 100 grammes of the solution. (c) The weight of solute dissolved in a litre of the solvent. (d) The weight of solute in a litre of the solution. The most usual method is to give the concentration as the weight of solute dissolved in 100 grammes or in one litre of the solvent.

Molecular Concentration.--Many of the physical and biological properties of a solution are proportional, not to its mass or weight concentration, but to its molecular concentration, i.e. to the number of gramme-molecules of the solute contained in a litre of the solution. Many physical properties are quite independent of the nature of the solute, depending only on its degree of molecular concentration.

Normal Solution.--A normal solution is one which contains one gramme-molecule of the solute per litre. A decinormal solution contains one-tenth of a gramme-molecule of the solute per litre, and a centinormal solution one-hundredth of a gramme-molecule. A normal solution of urea, for example, {16} contains 60 grammes of urea per litre, while a normal solution of sugar contains 342 grammes of sugar per litre.

The Dissolved Substance is a Gas.--Van t' Hoff, using the data obtained by the botanist Pfeffer, showed that the dissolved matter in a solution behaved exactly as if it were a gas. The analogy is complete in every respect. Like the gaseous molecules, the molecules of a solute are mobile with respect to one another. Like those of a gas, the molecules of a solute tend to spread themselves equally, and to fill the whole space at their disposal, i.e. the whole volume of the solution. The surface of the solution represents the vessel containing the gas, which confines it within definite limits and prevents further expansion.

Osmotic Pressure.--Like the molecules of a gas, the molecules of a solute exercise pressure on the boundaries of the space containing it. This osmotic pressure follows exactly the same laws as gaseous pressure. It has the same constants, and all the notions acquired by the study of gaseous pressure are applicable to osmotic pressure. Osmotic pressure is in fact the gaseous pressure of the molecules of the solute.

When a gas dilates and increases in volume, its temperature falls, and cold is produced. Similarly, when a soluble substance is dissolved, it increases in volume, and the temperature of the liquid falls. This phenomenon is well known as a means of producing cold by a refrigerating mixture.

The phenomena of life are governed by the laws of gaseous pressure, since all these phenomena take place in solutions. The fundamental laws of biology are those of the distribution of substances in solution, which is regulated by the laws of gaseous pressure, since all these laws are applicable also to osmotic pressure.

Boyle's Law.--When a gas is compressed its volume is diminished. If the pressure is doubled, the volume is reduced to one-half. The quantity V ?P, that is the volume multiplied by the pressure, is constant.

Gay-Lussac's Law.--For a difference of temperature of a degree Centigrade all gases dilate or contract by $1 / 273$ of their volume at 0?Centigrade. {17}

Dalton's Law.--In a gaseous mixture, the total pressure is equal to the sum of the pressures which each gas would exert if it alone filled the whole of the receptacle.

Pressure proportional to Molecular Concentration.--The above laws are completely independent of the chemical nature of the gas, they depend only on the number of gaseous molecules in a given space, i.e. on the molecular concentration. If we double the mass of the gas in a given space, we double the number of molecules, and we also double the pressure, whatever the nature of the molecules. We may also double the pressure by compressing the molecules of a gas, or of several gases, into a space half the original size. The molecular concentration of a gas, or of a mixture of gases, is the ratio of the number of molecules to the volume they occupy. The pressure of a gas or of a mixture of gases is proportional to its molecular concentration. This is a better and a shorter way of expressing both Boyle's law and Dalton's law.

One gramme-molecule of a gas, whatever its nature, condensed into the volume of 1 litre, has a pressure of 22.35 atmospheres. Similarly one gramme-molecule of a solute, whatever its nature, when dissolved in a litre of water, has the same pressure, viz. 22.35 atmospheres.

Absolute Zero.--According to Gay-Lussac's law, the volume of a gas diminishes by 1 / 273 of its volume at 0°C. for each degree fall of temperature. Thus if the contraction is the same for all temperatures, the volume would be reduced to zero at -273°C. This is the absolute zero of temperature. Temperatures measured from this point are called absolute temperatures, and are designated by the symbol T. If t° indicates the Centigrade temperature above the freezing point of water, then the absolute temperature is equal to t° + 273°

The Gaseous Constant.--Consider a mass of gas at 0°C. under a pressure Po, with volume Vo. At the absolute temperature T, if the pressure be unaltered, the volume of this gas will be VoT / 273. Therefore the constant PV, the product of the pressure by the volume, will be represented by PoVoT / 273. {18}

At the same temperature, but under another pressure P' the gas will have a different volume V'. Since, according to Boyle's law, PV is constant (P'V' = PoVo), it will still equal PoVoT / 273. Therefore PoVo / 273 is also constant. This quantity is called "the gaseous constant," and if we represent it by the symbol R, we obtain the general formula PV = RT for all gases, or PV / T = R.

Suppose, for instance, we have a gramme-molecule of a gas at 0°C. in a space of 1 litre. It has a pressure of 22.35 atmospheres at 0°C., or 273° absolute temperature. Since PV = RT, R = PV / T = 1 ×22.35 / 273 = .0819. This number .0819 is the numerical value of the constant R for all gases, volume being measured in litres and pressure in atmospheres.

Substances in solution behave exactly like gases, they follow the same laws and have the same constants. All the conceptions which have been acquired by the study of gases are applicable to solutions, and therefore to the phenomena of life. The osmotic pressure of a solution is the force with which the molecules of the solute, like gaseous molecules, strive to diffuse into space, and press on the limits which confine them, the containing vessel being represented by the surfaces of the solution. Osmotic pressure is measured in exactly the same way as gaseous pressure. To measure steam pressure we insert a manometer in the walls of the boiler. In the same way we may use a manometer to measure osmotic pressure. We attach the tube to the walls of the porous vessel, allow the solvent to increase in volume under the pressure of the solute, and measure the rise of the liquid in the manometer tube.

Pfeffer's Apparatus.--Pfeffer has designed an apparatus for the measurement of osmotic pressure. It consists of a vessel of porous porcelain, the pores of which are filled with a colloidal solution of ferrocyanide of copper. This forms a semi-permeable membrane which permits the passage of water into the vessel, but prevents the passage of sugar or of any {19} colloid. The stopper which hermetically closes the vessel is pierced for the reception of a mercury manometer. The vessel is filled with a solution of sugar and plunged in a bath of water. The volume of the solution in the interior of the vessel can vary, since water passes easily in either direction through the pores of the vessel. The boundary of the solvent has become extensible, and its volume can increase or diminish in accordance with the osmotic pressure of the solute. Under the pressure of the sugar water is sucked into the vessel like air into a bellows, the solution passes into the tube of the manometer, and raises the column of mercury until its pressure balances the osmotic pressure of the sugar molecules.

Osmotic Pressure follows the Laws of Gaseous Pressure.--This osmotic pressure is in fact gaseous pressure, and may be measured in millimetres of mercury in just the same way. We may thus show that osmotic pressure follows the laws of gaseous pressure as defined by Boyle, Dalton, and Gay-Lussac. The coefficient of pressure variation for change of temperature is the same for a solute as for a gas. The formula $PV = RT$ is applicable to both. The numerical value of the constant R is also the same for a solute as for a gas. being .0819 for one gramme-molecule of either, when the volume is expressed in litres and the pressure in atmospheres. The formula $PV = RT$ shows that for a given mass, with the same volume, the pressure increases in proportion to the absolute temperature.

Osmotic Pressure of Sugar.--A normal solution of sugar, containing 342 grammes of sugar per litre, has a pressure of 22.35 atmospheres, and it may well be asked why such an enormous pressure is not more evident. The reason will be found in the immense frictional resistance to diffusion. Frictional resistance is proportional to the area of the surfaces in contact, and this area increases rapidly with each division of the substance. When a solute is resolved into its component molecules, its surface is enormously increased, and therefore the friction between the molecules of the solute and those of the solvent.

Isotonic Solutions.--Two solutions which have the same {20} osmotic pressure are said to be iso-osmotic or isotonic. When comparing two solutions of different concentration, the solution with the higher osmotic pressure is said to be hypertonic, and that with the lower osmotic pressure hypotonic.

Lowering of the Freezing Point.--Pure water freezes at 0?C. Raoult showed that the introduction of a non-ionizable substance, such as sugar or alcohol, lowers the freezing point of a solution in proportion to the molecular concentration of the solute. One gramme-molecule of the solute introduced into one litre of the solution lowers its temperature of congelation by 1.85?C. Thus a normal solution of any non-ionizable substance in water freezes at - 1.85?C. The measurement of this lowering of the freezing point is called Cryoscopy, a method which is becoming of great utility in medicine.

Cryoscopy of Blood.--In order to determine the osmotic pressure of the blood at 37?C., i.e. 98.6?F., the normal temperature, we proceed as follows. On freezing the blood, we find that it congeals at -.56? Its molecular concentration is therefore .56 / 1.85 = .30, or about one-third of a gramme-molecule per litre. Its osmotic pressure at 0?C. is therefore .3 ?22.35 = 6.7 atmospheres. The increase of pressure with temperature is the same as for a gas, viz. 1/273, or .00367 of its pressure at 0?for every degree rise of temperature. The increase of pressure at 37?is therefore .00367 ?37 ?6.7 = .9 atmospheres. The total osmotic pressure at 37?is therefore 6.7 + .9 = 7.6 atmospheres.

Rise of Boiling Point.--Water under atmospheric pressure boils at a temperature of 100?C. The addition of a solute whose solution does not conduct electricity, such as sugar, causes a rise in the boiling point proportional to the molecular concentration of that solute.

Lowering of the Vapour Tension.--The vapour tension of a liquid is lowered by the addition of a solute. A liquid boils at the temperature at which its vapour tension equals that of the atmosphere. Since an aqueous solution of sugar at atmospheric pressure does not begin to boil at 100?C., it is manifest that its vapour tension is then less than that of the {21} atmosphere. The addition of a solute such as sugar, whose solution is not ionizable, and therefore does not conduct electricity, lowers the vapour tension of the solution in proportion to the molecular concentration of the solute.

Corresponding Values.--We have thus found five properties of a solution which vary proportionally, so that from the measurement of any one of them we can determine the corresponding values of all the others. These are--

1. The Molecular Concentration. 2. The Osmotic Pressure. 3. The Diminution of Vapour Tension. 4. The Raising of the Boiling Point. 5. The Lowering of the Freezing Point.

Cryoscopy.--The usual method employed for the determination of the molecular concentration and osmotic pressure of a solution is by cryoscopy-- the measurement of its temperature of congelation. A very sensitive thermometer is used, the scale of which extends over only 5?and is divided into hundredths of a degree. The liquid under examination is placed in a test tube, in which the bulb of the thermometer is plunged, and this is supported in a second tube with an air space all round it. The whole is then suspended to the under side of the cover of the refrigerating vessel, which may be cooled either by filling it with a freezing mixture, or by the evaporation of ether. During the whole of the operation the liquid is agitated by a mechanical stirrer. The first step is to determine the freezing point of distilled water. As the water cools the mercury gradually descends in the stem of the thermometer till it reaches a point below the zero mark at 0?C. As soon as ice begins to form the mercury rises, at first rapidly and then more slowly, reaches a maximum, and finally descends again. This maximum reading is the true point of congelation. The inner tube is then emptied, care being taken to leave a few small ice crystals to serve as centres of congelation for the subsequent experiment, thus avoiding supercooling of the solution. The process is then repeated with the solution under examination. The difference between {22} the two freezing points is the required "lowering of the freezing point."

Cryoscopy is the method most used in biological research to determine molecular concentration. It has, however, some grave defects. It necessitates several cubic centimetres of the liquid under examination. It gives us the constants of the solution at the temperature of freezing, which is far below that of life. Organic liquids are easily altered and are extremely sensible to minute differences of temperature, cryoscopy therefore gives us no information as to the constitution of solutions under normal conditions. It is desirable to have some other method of determining molecular concentration and the other interdependent constants at the normal temperature of life. A much better method, were it possible, would be the direct determination of the vapour tension of the solutions under normal conditions of temperature and pressure.

Molecular Lowering of the Freezing Point.--For every substance whose solution is not ionized and therefore does not conduct electricity, the lowering of the freezing point is the same, viz. 1.85?C. for each gramme-molecule of the solute per litre of the solution.

Determination of the Molecular Concentration.--In order to obtain the molecular concentration of a non-ionizable substance, we have only to determine the lowering of the freezing point. Let A be the lowering of the freezing point of any solution. On dividing it by 1.85 (the lowering of the freezing point for a normal solution), we obtain the number of gramme-molecules in a litre of the solution. If n be the number of gramme-molecules per litre, then $n = A / 1.85$.

Determination of the Osmotic Pressure.--The osmotic pressure P of a solution may be obtained by multiplying its molecular concentration n by 22.35 atmospheres. $P = n ?22.35 = A / 1.85 ?22.35$.

Determination of Molecular Weight.--The lowering of the freezing point also enables us to calculate the molecular {23} weight of any non-ionizable solute. Thus Bouchard has been able to determine by means of cryoscopy the mean molecular weight of the substances eliminated by the urine. A weight x of the substance is dissolved in a litre of water, and the lowering of the freezing point is observed. The value thus found divided by 1.85 gives us n, the number of gramme-molecules per litre. The molecular weight M may be determined by dividing the original weight x by n.

The study of osmotic pressure was begun by the Abb?Nollet; and one of his disciples, Parrot, at an early date thus described its importance: "It is a force analogous in all respects to the mechanical forces, a force able to set matter in motion, or to act as a static force in producing pressure. It is this force which causes the circulation of heterogeneous matter in the liquids which serve as its vehicle. It is this force which produces those actions which escape our notice by their minuteness and bewilder us by their results. It is for the infinitely small particles of matter what gravitation is for heavy masses. It can displace matter in solution upwards against gravity as easily as downwards or in a horizontal direction."

Thus the recognition of the fact that a substance in solution is really a gas, has at a single stroke put us in possession of the laws of osmotic pressure-- laws slowly and laboriously discovered by the long series of investigations on the pressure of gases.

Osmotic pressure plays a most important role in the arena of life. It is found at work in all the phenomena of life. When osmotic pressure fails, life itself ceases.

* * * * *

{24}

CHAPTER III

ELECTROLYTIC SOLUTIONS

Solutions which conduct Electricity.--The laws of solution which we have studied in the previous chapter apply only to those solutions, chiefly of organic origin, which do not conduct electricity. Solutions of electrolytes such as the ordinary salts, acids, and bases, which are ionized on solution, give values for the various constants of solution which do not accord with those required by theory. If, for instance, we take a gramme-molecule of an electrolyte such as chloride of sodium, and dissolve it in a litre of water, we find that the lowering of the freezing point is nearly double the theoretical value of 1.85? The same holds good for the osmotic pressure, and for all the constants which are proportional to the molecular concentration of the solute. The solution behaves, in each case, as if it contained more than one gramme-molecule of sodium chloride per litre. It behaves, in fact, as if it contained i times the number of molecules of solute originally introduced into it. If n be the original number of molecules, then it will apparently contain n' = in molecules. This law is universal for all electrolytic solutions; the theoretical value for their concentration, osmotic pressure, and all the proportional physical constants must be multiplied by this quantity, i = n'/n, which is the ratio of the apparent number of the molecules present to the number originally introduced.

A similar dissociation of the molecule is observed in the case of many gases. The vapour of chloride of ammonium, for instance, is decomposed by heat, and it may be shown experimentally that the increase of pressure on heating above {25} that which theory demands, is due to an increase in the number of the gaseous molecules present. Some of the vapour particles are dissociated into two or more fragments, each of which plays the part of a single molecule.

Arrhenius, in 1885, advanced the hypothesis that the apparent increase in the number of molecules of an electrolytic solution was also due to dissociation. This interpretation at once threw a flood of light on a number of phenomena hitherto obscure.

Coefficient of Dissociation.--We have seen that in order to obtain values which accord with experiment we have to multiply the number of gramme-molecules of the solute by the coefficient i, which is called the Coefficient of Dissociation.

This coefficient of dissociation, i, may be found by observing the lowering of the freezing point of a normal solution, and dividing it by 1.85. i = t/1.85.

The coefficient of dissociation varies with the degree of concentration of the solution, rising to a maximum when the solution is sufficiently diluted.

If we know i, the coefficient of dissociation for a given solute, contained in a solution of a definite concentration, we can find n', the number of particles present in a solution containing n gramme-molecules of the solute per litre, since n' = in. On the other hand, if from a consideration of its freezing point and other constants we find that an electrolytic solution appears to contain n' gramme-molecules per litre, the real number of chemical gramme-molecules in one litre of the solution will be only n' / i = n.

Very concentrated solutions do not conform to these laws. In this they resemble gases, which as they approach their point of condensation tend less and less to conform to the laws of gaseous pressure.

Electrolysis.--If we take a solution of an acid, a salt, or a base, and dip into it two metallic rods, one connected to the positive and the other to the negative pole of a battery, we {26} find that the metals or metallic radicals of the solution are liberated at the negative pole, while the acid radicals of the salts and acids and the hydroxyl of the bases are liberated at the positive pole. The liberated substances may either be discharged unchanged, or they may enter into new combinations, causing a series of secondary reactions.

Electrolytes.--Solutions which conduct electricity are called Electrolytes, and the conducting metallic rods dipping into the solution are the Electrodes. Faraday gave the names of Ions to the atoms or atom-groups liberated at either electrode. The ions liberated at the positive electrode are the Anions, and those at the negative electrode are the Cations. The only solutions which possess any notable degree of electrical conductivity are the aqueous solutions of the various salts, acids, and bases, and in these solutions only do we meet with those phenomena of dissociation which are evidenced by anomalies of osmotic pressure, freezing point and the like,--anomalies which show that the solution contains a greater number of molecules than that indicated by its molecular concentration. These anomalies are due to dissociation, the division of some of the molecules into fragments, each of which plays the part of a separate molecule, contributing its quota to the osmotic tension and vapour pressure of the solution, in fact to all the phenomena which are dependent on the degree of molecular concentration. The electrical conductivity of a solution is therefore proved to be dependent on its molecular dissociation.

Arrhenius' Theory of Electrolysis.--In 1885, Arrhenius brought forward his theory of the transport of electricity by an electrolyte. According to this hypothesis, the electric current is carried by the ions, the positive charges by the cations, and the negative charges by the anions. In virtue of the attraction between charges of different sign, and repulsion between charges of like sign, the cations are repelled by the positive charge on the anode, and attracted by the negative charge on the cathode. Similarly the anions are repelled by the cathode and attracted by the anode. {27}

An electrolytic solution contains three varieties of particles, positive ions or cations, negative ions or anions, and undissociated neutral molecules. The molecular concentration of such a solution, with the corresponding constants, depends on the total number of these particles, i.e. the sum of the ions and the undissociated neutral molecules. We may indicate an ion by placing above it the sign of its electrical charge, one sign for each valency. Thus Na^+ and Cl^- indicate the two ions of a salt solution; Cu^{++} and $SO4^{--}$ the two ions of a solution of sulphate of copper. A point is sometimes substituted for the + sign, and a comma for the - sign. Thus $Na^.$ and $Cl^,$; $Cu^{..}$ and $SO4^{,,}$.

My friend Dr. Lewis Jones has given a very vivid picture of the processes which go on in an electrolytic solution when an electric current is passing. He compares an electrolytic cell to a ballroom, in which are gyrating a number of dancing couples, representing the neutral molecules, and a number of isolated ladies and gentlemen representing the anions and cations respectively. If we suppose a mirror at one end of the ballroom and a buffet at the other, the ladies will gradually accumulate around the mirror, and the gentlemen around the buffet. Moreover, the dancing couples will gradually be dissociated in order to follow this movement.

Degree of Dissociation.--The degree of dissociation is the fraction of the molecules in the solution which have undergone dissociation. Let n be the total number of molecules of the solute, and n" the number of dissociated molecules. Then n" / n = a will represent the degree of dissociation. Let k be the number of ions into which each molecule is split. Then a = n"k / nk, i.e. the degree of dissociation is the ratio of the number of ions actually present in a solution to the number which would be present if all the molecules of the solute were dissociated.

Let n' be the total number of particles present in a solution {28} containing n molecules, each of which is composed of k ions. Then if a is the degree of dissociation,

$$n' = n - an + ank, \quad n' = n[1 + a(k-1)], \quad n'/n = 1 + a(k-1) = i.$$

We thus obtain i the coefficient of dissociation, in terms of the degree of dissociation a and the number of ions in each molecule k.

If there is no dissociation, i.e. if a = 0, then n' = n, and i = 1. If all the molecules are dissociated, a = 1, and i = k.

Faraday's Law.--Faraday found that the quantity of electricity required to liberate one gramme-molecule of any radical is 96.537 coulombs for each valency of the radical.

Electrochemical Equivalent.--The electrochemical equivalent of a radical is the weight liberated by one coulomb of electricity. It is equal to the molecular weight of the ion, divided by 96.537 times its valency.

Electrolytic Conductivity.--The conductivity of an electrolyte is the inverse of its resistance. C = 1/R.

For a given difference of potential the conductivity of an electrolyte is proportional to the number of ions in unit volume, the electrical charge on each ion, and the velocity of the ions.

The specific conductivity [Delta] of an electrolyte is the conductivity of a cube of the solution, each face of which is one square centimetre in area. The molecular conductivity of an electrolyte is the conductivity of a solution containing one gramme-molecule of the substance placed between two parallel conducting plates, one centimetre apart. The molecular conductivity is independent of the volume occupied by the gramme-molecule of the solute, depending only on the degree of dissociation. The molecular conductivity U is equal to the product of V, the volume of the molecule, by [Delta], its specific conductivity. U = V[Delta]. Whence [Delta] = U / V, i.e. the specific {29} conductivity equals the molecular conductivity divided by the volume.

The conductivity of an electrolyte is proportional to the number of ions in a volume of the solution containing one gramme-molecule. Let M{[infinity]} be the conductivity for complete dissociation and Mv the molecular conductivity at the volume V. Then

$$Mv / M\{[\text{infinity}]\} = n''k / nk = n'' / n = a,$$

the degree of dissociation. This is Ostwald's law, which says that the degree of dissociation is equal to the ratio of conductivity when the gramme-molecule occupies a volume V, to its conductivity when the solution is so dilute that dissociation is complete. Hence the degree of dissociation may also be determined by comparing the electrical conductivities of two solutions of different degrees of concentration.

```
| -- -- -- | -- -- -- | | SO4 SO4 SO4 | SO4 SO4 SO4 | | | | | ++ ++ ++ | ++ ++ ++
| | Cu Cu Cu | Cu Cu Cu | | | | +---------------------------+----------------------------
-+
```

FIG. 1.--Before the passage of the current.

```
| -- | -- -- | | SO4 | SO4 SO4 SO4 SO4 SO4 | - | | | + | ++ | ++ ++ | | Cu Cu Cu
Cu | Cu Cu | | | | +-------------------------+-------------------------------+
```

FIG. 2.--After the passage of the current.

Velocity of the Ions.--If the electrolytic cell is divided into two segments by means of a porous diaphragm, we shall find after a time an unequal distribution of the solute on the two sides. For instance, with a solution of sulphate of copper, after the current has passed for some time there will be a diminution of concentration in the liquid on both sides of the diaphragm, but the loss will be very unequally divided. Two-thirds of the loss of concentration will be on the side of the negative electrode and only one-third on the positive side. In 1853, Hittorf gave the following ingenious explanation of this phenomenon:-- {30}

Fig. 1 represents an electrolytic vessel containing a solution of sulphate of copper, the vertical line indicating a porous partition separating the vessel into two parts. Fig. 2 shows the same vessel after the passage of the current. The acid radical has travelled twice as fast as the metal. For each copper ion which has passed through the porous plate towards the cathode two acid radicals have passed through it towards the anode. Three ions have been liberated at either electrode, but in consequence of the difference of velocity with which the positive and the negative ions have travelled, the negative side of the vessel contains only one molecule of copper sulphate and has lost two-thirds of its molecular concentration, while the positive side contains two molecules of copper sulphate and has only lost one-third of its concentration. This proves clearly that the ions move in different directions with different velocities. Let u be the velocity of the anions, and v the velocity of the cations. Let n be the loss of concentration at the cathode, and 1 - n the loss of concentration at the anode. Then

$$u / v = n / (1 - n),$$

i.e. the loss of concentration at the cathode is to the loss of concentration at the anode as the velocity of the anions is to that of the cations. Hence by measuring the loss of concentration at the two electrodes, we have an easy means of determining the comparative velocity of different ions.

In 1876, Kohlrausch compared the conductivity of the chlorides, bromides, and iodides of potassium, sodium, and ammonium respectively. He found that altering the cation did not affect the differences of conductivity between the three salts, thus showing that these differences of conductivity were dependent on the nature of the anion only, and not on the particular base with which it was combined. The difference of conductivity between an iodide and a bromide, for example, is the same whether potassium, sodium, or ammonium salts are compared. A similar experiment has been made with a series of cations combined with various anions. The difference of conductivity of the salts in the series is the same whichever anion is used, i.e. the difference of conductivity between potassium chloride and sodium chloride is the same as that between {31} potassium bromide and sodium bromide. Hence we may conclude that the conductivity of any salt is an ionic property.

Kohlrausch's law may be expressed by the formula $c = d(u + v)$, where c is the conductivity of the salt, d the degree of dissociation, i.e. the fraction of the electrolyte broken up into ions, and u and v the velocity of the anions and cations respectively. When all the molecules of the electrolyte are dissociated, $d = 1$, and the formula becomes $c_{[\infty]} = u + v$.

As we have already seen, a salt is formed by the union of a metal M with an acid radical R. Potassium sulphate, K_2SO_4, consists of the metal K_2 and the acid radical SO4. Ammonium chloride, NH4Cl, consists of the basic radical NH4 and the acid radical Cl. The various acids may be considered as salts of the metal hydrogen. Thus sulphuric acid, H2SO4, is the sulphate of hydrogen. Bases may be considered as salts with the hydroxyl group, OH, replacing the acid radical. Thus potash, KOH, is the hydroxyl of potassium. The various electrolytic combinations may be represented by the following symbols:--

Salts = MR. Acids = HR. Bases = MOH.

The various chemical reactions of an electrolyte are all ionic reactions, the chemical activity of an electrolytic solution being proportional to its electric conductivity, i.e. the degree of dissociation of its ions. The acidity of an electrolytic solution is due to the presence of the dissociated ion H^+, and its strength is determined by the concentration of these free hydrogen ions. Hence the greater the degree of dissociation the stronger the acid.

The basic character of a solution is determined by the presence of the hydroxyl radical OH^-. The greater the concentration of the hydroxyl ions, i.e. the greater the dissociation, the stronger is the base.

The ions H^+ and OH^- are of special importance, since they are the ions of water, $H2O = H^+ + OH^-$. The degree of {32} dissociation of pure water is but small. Water is, however, the most important of all the various agents in the chemical reactions of life, since a large number of organic substances are decomposed by water by a process of hydrolysis, and a vast number of organic substances are but combinations of carbon with the ions H^+ and OH^-, their diversity being due to variations in the relative proportions and grouping.

The Chemical, Therapeutic, and Toxic Actions of Ions.--The chemical, therapeutic, antiseptic, and toxic actions of electrolytic solutions are almost exclusively due to ionization. Take, for instance, a solution of nitrate of silver in which the addition of chlorine produces a white precipitate of chloride of silver. This precipitate occurs only when the solution added is one such as NaCl, where the chlorine is present as the free ion Cl^-. No such precipitate is produced in a solution of chlorate of potassium or chloracetic acid, where the chlorine is entangled in the complex ion ClO3 or C2H3ClO2.

Since, then, the toxic and pharmacological properties of an electrolyte depend entirely on the ionic grouping, it behoves the physician and the biologist to study the structure and grouping of the ions in a molecule, rather than that of the atoms. Consider for a moment the totally different properties of the phosphides and the phosphates. The former are extremely toxic, while the latter are perfectly harmless. There is not the slightest analogy between their actions on the living organism. On the other hand, all the phosphides produce the same toxic and therapeutic effects, whatever the cation with which they are united. Their toxic properties are derived from the presence of the free phosphorus ion P^{---}. The phosphates contain phosphorus in the same proportion as the phosphides, but this phosphorus is harmlessly entangled in the complex ion $PO4^{---}$, whose properties are absolutely different from those of the ion P^{---}.

The above considerations apply equally to the chlorides and chlorates, the iodides and iodates, the sulphides and sulphates, and in general to all chemical salts. {33}

The question has an intimate bearing on practical pharmacology. When we prescribe a cacodylate or an amylarsinate, we are not prescribing an arsenical treatment whose effects can be compared with those of an arsenide, an arsenite, or an arsenate. This fact is sufficiently indicated by the difference in the toxic doses of the different salts. Each variety of arsenical ion has its own special physiological and therapeutic properties. We do not expect to obtain the results of a ferruginous treatment from the administration of a ferrocyanide or a ferricyanide. Both contain iron, it is true, but neither possess the properties of the cation Fe^{+++}, but rather those of the complex anion of which they form a part.

We have already said that most of the therapeutic, toxic, and caustic actions of an electrolyte are due to ionic action, and the substances can therefore have no toxic action unless they are dissociated. Many of the solvents employed in medicine, such as alcohol, glycerine, vaseline, and chloroform dissolve the electrolytes but do not dissociate them into ions, and these solutions therefore do not conduct electricity. Such solutions have no therapeutic action. With the absence of dissociation all the ionic toxic and caustic effects also disappear entirely, and only re-appear as the water of the tissue is able slowly to effect the necessary dissociation.

Carbolic acid dissolved in glycerine is hardly caustic and but very slightly toxic. We have met with several instances in which a tablespoonful of carbolized glycerine, in equal parts, has been swallowed without any ill effect, either caustic or toxic, whereas the same dose dissolved in water would have been fatal. This absence of dissociation has enabled the surgeon Mencie to inject carbolic and glycerine in equal proportions into the larger joints, the part being subsequently washed out with pure alcohol. Thus by employing vaseline, oil, or glycerine as a solvent, and avoiding the access of water, we are able to use electrolytic antiseptics in very concentrated form. Their action is brought out very slowly, as the water of the organism effects the necessary dissociation of the electrolyte. {34}

Since all chemical, toxic, and therapeutic actions are ionic, they are proportional to the degree of ionic concentration, i.e. to the number of ions in a given volume. The only point of importance, that which determines their activity, whether chemical or therapeutic, is the degree of ionization or dissociation. For example, all acids have the same cation H^+. They have all identical properties, but they differ widely in the intensity of their action. There are weak acids such as acetic acid, and strong acids like sulphuric acid. The stronger acids are those which are more thoroughly dissociated, and in which the ion H^+ is very concentrated; whereas the feeble acids are but slightly dissociated, so that the ion H^+ is less concentrated.

Paul and Krig have shown that the bactericidal action of different salts also varies with their degree of dissociation, i.e. with the concentration of the active ions. They made a series of observations on the bactericidal action of various salts of mercury, the bichloride, the bibromide, and the bicyanide, on the spores of Bacillus anthracis. The following results were obtained from a comparison of solutions containing 1 gramme-molecule of the salt in 64 litres of water. With the bichloride solution, after exposure to the solution for twenty minutes, only 7 colonies of the bacillus were developed. After exposure to a similar solution of the bibromide the number of colonies was 34. The antiseptic action of the bichloride was therefore five times as great as that of the bibromide. The bicyanide of mercury, however, even when four times as concentrated, permitted the growth of an enormous number of colonies, showing that it had no appreciable antiseptic action whatever. Nevertheless, the proportion of Hg is the same in all the solutions, and if there were any difference one would naturally expect that the ion Cy^- would be more toxic than Cl^- or Br^-. The real condition which varies in these solutions and determines their activity is the degree of dissociation. The whole of the antiseptic property resides in the ion Hg^{++}. This ion is very {35} concentrated in the highly dissociated solution $HgCl_2$, less concentrated in the less ionized solution $HgBr_2$, and exceedingly dilute in the $HgCy_2$, which is hardly ionized at all.

What is true of the bactericidal action of the salts of mercury is equally true of their therapeutic effect. It is a great mistake to estimate the medicinal activity of a solution of a salt of mercury, or indeed of any electrolytic solution, simply by its degree of molecular concentration. The important point is the degree of dissociation, which is the only true measure of its activity. In the intramuscular injection of mercury salts it is by no means a matter of indifference what salt we employ. A salt should be used such as the bichloride or the biniodide, which is easily dissociated. Other salts are often employed because they occasion less pain at the site of injection; but the pain is a sign of the degree of activity of the preparation. The pain, it is true, may be avoided by using a salt which is less easily dissociated, or in which the mercury is bound up in a complex ion, but by so doing we diminish the efficacy of the remedy. It is moreover quite easy to diminish, or even entirely to suppress, the pain, by using a very dilute solution of an active ionized salt. A one-half per cent. or even one-quarter per cent. solution of the bichloride or biniodide of mercury may be injected very slowly in sufficient quantity without producing the slightest discomfort. Local action depends entirely on ionic concentration. One drop of pure sulphuric acid will destroy the skin, whereas the same amount if diluted in a tumblerful of water will furnish a refreshing drink.

* * * * *

{36}

CHAPTER IV

COLLOIDS

As we have already seen, living organisms are formed essentially of liquids.

These liquids are solutions of crystallizable substances or crystalloids, and non-crystallizable substances or colloids--a classification which we owe to Graham.

The liquids are the most important constituents of a living organism, since they are the seat of all the chemical and physical phenomena of life. The junction of two liquids of different concentration is the arena in which takes place both the chemical transformation of matter and the correlative transformation of energy. In a former chapter we have passed in review the class of crystalloids, we will now turn our attention to the characteristic properties of colloids.

Colloids.--Colloids differ from crystalloids in that they do not form crystals from solution, being completely amorphous when in the solid state. The solution of a colloid solidifies in the same form which it possessed in the liquid state, the solvent being enclosed in the meshes of a sort of network formed by the solute. This form is approximately retained even after the water has evaporated by drying, the passage from the liquid state of solution to the solid state being effected through a series of intermediary states, such as a clot, coagulum, or jelly. This passage from the state of solution into a state of jelly is called coagulation. Some colloids, such as gelatine, coagulate with cold; while others, such as egg-albumin, coagulate with heat. Some, like the caseine of milk, require the addition of certain chemical substances to set up coagulation; while still others, such as the fibrin of blood, appear to coagulate spontaneously. The physical phenomena of {37} coagulation are still but little understood. In some cases it is a reversible phenomenon, thus gelatine coagulated by cold is redissolved by heat; whereas with other colloids the process is irreversible, albumin coagulated by heat is not redissolved on cooling.

Colloids in a state of coagulation have a vacuolar or sponge-like structure. The solvent is imprisoned in the vacuoles of the clot, and is expelled little by little by its retraction. Colloids diffused in water are usually called colloidal solutions, but they are not true solutions. Such a pseudo-solution of a colloid is called a "sol," while a colloid in a state of coagulation is called a "gel." Colloidal solutions spread but little, diffuse very slowly in the liquids of the body, and cannot penetrate organic membranes.

Colloidal solutions diffuse light, unlike crystalloid solutions, which are transparent. We all know how the trajectory of a beam of sunlight through a darkened room is rendered visible by the particles of dust. In the same way if a colloidal solution is illuminated by a transverse ray of light, the light is diffused by the molecules of the colloid in semi-solution, and the liquid appears faintly illuminated on a dark background. The light diffused by a colloidal solution is polarized, which shows that it is reflected light,

Siedentopf and Sigmondy have applied this principle of lateral illumination on a dark background to the construction of the ultra-microscope. With the aid of this instrument we may not only see, but count the particles in a colloidal solution, which is in reality merely a pseudo-solution or suspension, in contradistinction to the true solution of a crystalloid.

Colloidal solutions possess only a very feeble osmotic pressure. The lowering of the freezing point and the other corresponding constants are also quite insignificant. This arises from the fact that the molecules of a colloid are extremely large when compared with those of a crystalloid. For example let us take colloidal substance whose molecular weight is 2000. A solution containing 40 grammes per litre would have an osmotic pressure only one-fiftieth of that of a {38} solution of similar strength of a crystalloid whose molecular weight was 40.

Not only so, but on measuring the molecular concentration, the osmotic pressure, and the other constants of a colloidal solution, we find values even lower than those which we should expect from a consideration of its molecular weight. This is probably due to the tendency of a colloid to polymerization, i.e. to form groups or associations of molecules. Suppose, for instance, that the molecules of a colloidal solution are aggregated into groups of ten. Since each group plays the part of a simple molecule, the osmotic pressure will be ten times less than that corresponding to the quantity of the solute present. Such a group of molecules is called by Naegeli a "micella."

Similar phenomena of aggregation may be observed in the molecules of many inorganic substances. The molecule of iodine, for example, is monatomic at 1200?C., but becomes diatomic at the ordinary temperature. Sulphur at 860?C. is a gas with a vapour density of 2.2, while at 500?C. its vapour density rises to 6.6. In both of these cases two or more molecules of the element have been condensed into one as a result of the fall of temperature.

We frequently find that two successive cryoscopic observations on the freezing point of the same colloidal solution will vary. This is due to the extreme sensitiveness of the micell? which absorb or abandon their extra molecules under the slightest influence. This mobility in the constitution of the micell?appears to be one of the principal causes of the peculiar properties of colloidal solutions.

The phenomenon of polymerization appears to be reversible. The micell?are formed under certain conditions, and are disintegrated when these conditions are removed. The osmotic pressure varies in the same manner, diminishing with polymerization and augmenting with the disintegration of the micell? One may easily understand what an important role is played by this alternate polymerization and disintegration in the phenomena of life.

Most colloidal substances are precipitated from their solutions by the addition of very small quantities of electrolytic {39} solutions. Non-electrolytic solutions do not appear to provoke this precipitation. This is not a chemical action, for an exceedingly small quantity of an electrolyte is able to precipitate an indefinite quantity of the colloid. The precipitation is probably due to the electric charges carried by the dissociated ions of the electrolytes.

When an electric current is passed through a colloid solution, the course of the molecules of the colloid is sometimes towards the cathode and sometimes towards the anode, according to the nature of the colloid and of the solvent. This displacement would appear to indicate a difference of electric potential between the molecules of the colloid and those of the solvent. Hardy has shown that in an alkaline solution the molecules of albumin travel towards the anode, while in an acid solution they travel towards the cathode.

Metallic Colloids.--Carey Lea and afterwards Cred?succeeded in obtaining silver in colloidal solution by ordinary chemical means. Professor Bredig has introduced a more general method of obtaining a number of metals in colloidal solutions in a state of great purity. He causes an electric arc to pass between two rods of the metal immersed in distilled water. The cathode is thus pulverized into a very fine powder which rests in suspension in the liquid, constituting a colloidal solution. Bredig has in this way prepared sols of platinum, palladium, iridium, silver, and cadmium.

Catalytic Properties of Colloids.--Catalysis is the property possessed by certain bodies of initiating chemical reaction. The mass of the catalyzing body has no definite proportion to that of the substances entering into the reaction, and the appearance of the catalyzer is in no way altered by the reaction.

Ostwald has shown that catalysis consists essentially in the acceleration or retardation of chemical reactions which would take place without the action of the catalyzer, but more slowly.

Catalytic reactions are very numerous in chemistry. The inversion of sugar by acids, the etherization of alcohol by sulphuric acid, the decomposition of hydrogen peroxide by {40} platinum black are all instances of catalysis. Fermentation by means of a soluble ferment or diastase, a phenomenon which may almost be called vital, is also a catalytic action. The action of pepsin, of the pancreatic ferment, of zymase, and of other similar ferments has a great analogy with the purely physical phenomenon of catalysis. The diastases are all colloids, and so are many other catalyzers.

A catalyzer is a stimulus which excites a transformation of energy. The catalyzer plays the same role in a chemical transformation as does the minimal exciting force which sets free the accumulation of potential energy previous to its transformation into kinetic energy. A catalyzer is the friction of the match which sets free the chemical energy of the powder magazine.

Bredig has studied the catalytic decomposition of hydrogen peroxide by metallic colloids prepared by his electric method. He found that 1 atom-gramme of colloidal platinum gives a sensible catalytic effect when diluted with 70 million litres of water. Caustic soda and other chemical substances inhibit the catalytic action of colloidal platinum in the same way as they inhibit the fermenting action of diastase. The curve of decomposition of hydrogen peroxide by colloidal platinum may be compared with the curve of fermentation by emulsin. Both are equally affected by the addition of an alkali. Many other chemical and physical agents have a similar inhibitory action on the catalysis of colloidal metals and on diastasic fermentation. Thus a mere trace of sulphuretted hydrogen or hydrocyanic acid will paralyse the action of a colloidal metal, just as it does that of a ferment. This is what Bredig calls the poisoning of metallic ferments.

We may hope that the further study of catalysis, a purely physico-chemical phenomenon, may throw more light on the mechanism of diastasic fermentation, which is essentially a vital reaction.

It must not be forgotten that all classification is artificial and arbitrary, and only to be used as long as it facilitates study. This observation is particularly applicable to the classification of substances into crystalloids and colloids. {41} There is no sharp line between the two groups, the passage is gradual, and it is impossible to say where one group ends and the other begins. Many colloids such as hemoglobin are crystallizable, and many crystallizable substances are coagulable. Many substances appear at one time in the crystalloid state and at another time in the colloidal state, so that instead of dividing substances into colloids and crystalloids, we should rather consider these expressions as denoting different phases assumed by the same substance.

In order to define clearly our various classes and divisions, we are apt to exaggerate slight differences of properties or composition. We say that colloids have no osmotic pressure, whereas in fact the osmotic pressure of the colloids though feeble plays a very important part in the phenomena of life.

So in other departments of science--a factor which is almost infinitesimal may yet exercise a vast influence on the results. It is by infinitesimal variations of pressure, a thousandth of a millimetre or less, that we obtain the various degrees of penetration in the Roentgen rays.

The division into solutions and pseudo-solutions or suspensions is also an arbitrary one. A true solution is also a suspension of the molecules of the solute. There is no essential difference between a solution and a suspension, but only a difference in the size of the molecules, or agglomerations of molecules, in one case so small as to be transparent, and in the other case just big enough to diffuse light. There are moreover many properties common to colloidal solutions and suspensions of fine powders, such as kaolin, mastic, charcoal, or Indian ink. These particles in suspension are precipitated by solutions of electrolytes in a manner similar to the coagulation of colloids.

The surface of every liquid is covered by a very thin layer, a sort of membrane slightly differentiated from the rest of the liquid. This membrane may be a chemical one, a pellicular precipitate like that which is formed by the contact of two membranogenous liquids. On the other hand, the membrane may not differ from the subjacent liquid in chemical composition, but only in physical properties. If we {42} consider the molecules in the middle of a liquid, each molecule is subjected to the cohesive attraction of molecules on every side, attractions which neutralize one another. At the surface of the liquid, however, there are quite other conditions of equilibrium. There each molecule is drawn downwards towards the centre of the liquid, and there is no compensating attraction in an opposite direction. The resultant pressure is normal to the surface of the liquid, and is mechanically equivalent to an elastic membrane which tends to diminish the surface, and hence the volume of the liquid. We may therefore regard this surface tension as acting the part of a veritable physical membrane.

There is a still further differentiation of the surface of a liquid. When the liquid is not a simple one, but complex as in a solution, we find that the concentration of the solute is greater at the surface than in the interior. This is the so-called phenomenon of "adsorption," which is another cause for the production of a physical membrane covering the surface of a liquid.

Substances in a colloidal state have a great tendency to form these chemical or physical membranes at the point of contact between the colloidal solute and the solvent. This is probably the reason why the coagulum of a colloidal liquid usually presents a vacuolar or spongy structure.

* * * * *

{43}

CHAPTER V

DIFFUSION AND OSMOSIS

Diffusion and Osmosis.--If we place a lump of sugar in the bottom of a glass of water, it will dissolve, and spread by slow degrees equally throughout the whole volume of the liquid. If we pour a concentrated solution of sulphate of copper into the bottom of a glass vessel, and carefully pour over it a layer of clear water, the liquids, at first sharply separated by their difference of density, will gradually mix, so as to form a solution having exactly the same composition in all parts of the jar. The process whereby the sugar and the copper sulphate spread uniformly through the whole mass of the liquid in opposition to gravity is called Diffusion. This diffusion of the solute is a phenomenon exactly analogous to the expansion of a gas. It is the expression of osmotic pressure, or rather of the difference of the osmotic pressure of the solute in different parts of the vessel. The molecules of the solute move from a place where the osmotic pressure is greater towards a position where the osmotic pressure is less. The water molecules on the other hand pass from positions where the osmotic pressure of the solute is less towards positions where it is greater. As a consequence of this double circulation the osmotic pressure tends to become equalized in all parts of the vessel.

Diffusion appears to be the fundamental physical phenomenon of life. It is going on continually in the tissues of all living beings, and a study of the laws of diffusion and osmosis is therefore absolutely necessary for a just conception of vital phenomena.

Coefficient of Diffusion.--The coefficient of diffusion has {44} been defined by Fick as the quantity of a solute which in one second traverses each square centimetre of the cross section of a column of liquid 1 centimetre long, between the opposite sides of which there is unit difference of concentration. Nernst in his definition substitutes "unit difference of osmotic pressure" for "unit difference of concentration."

Until recently it was generally believed that diffusion took place in colloids and plasmas just as in pure water. This is, however, by no means the case: the differences are considerable. When a solute is introduced into a colloidal solution, the greater the concentration of the colloid the slower will be the diffusion. This may be shown by a simple experiment. Several glass plates are prepared, by spreading on each a solution of gelatine of different concentration, to which a few drops of phenol phthalein have been added. If now a drop of an alkaline solution be placed on each plate, we can see that the drop diffuses more slowly through the more concentrated gelatine solution, since the presence of the alkali is rendered visible by the coloration of the phenol phthalein. A similar demonstration may be made by allowing drops of acid to diffuse through solutions of gelatine made slightly alkaline and coloured with phenol phthalein. In general, we find on experiment that when similar drops of any coloured or colouring solution are left for an equal time on plates of gelatine of different degrees of concentration, the greater the concentration of the gelatine the smaller will be the circle of coloration obtained.

We may show that the rapidity of diffusion diminishes as the gelatinous concentration increases, by another experiment. If we put side by side on our gelatine plate a drop of sulphate of copper and another of ferrocyanide of potassium, the point of contact of the two fluids will be sharply marked by a line of precipitate. We find that under similar conditions the time between the sowing of the drops and the formation of this line of precipitate is longer when the gelatine is more concentrated.

Osmosis.--In 1748, l'Abb?Nollet discovered that when a pig's bladder filled with alcohol was plunged into water, the {45} water passed into the bladder more rapidly than the alcohol passed out; the bladder became distended, the internal pressure increased, and the liquid spirted out when the bladder was pricked by a pin. This passage of certain substances in solution through an animal membrane is called Osmosis, and membranes which exhibit this property are called osmotic membranes.

Precipitated Membranes.--In 1867, Traube of Breslau discovered that osmotic membranes could be made artificially. Certain chemical precipitates such as copper ferrocyanide can form membranes having properties analogous to those of osmotic membranes. With these precipitated membranes Traube made a number of interesting experiments. These have lately been collected in the volume of his memoirs published by his son.

Osmotic Membranes.--Osmotic membranes were formerly called semi-permeable membranes, being regarded as membranes which allow water to pass through them, but arrest the passage of the solute. This definition is inexact, since no membrane permeable to water is absolutely impermeable to the solutes. All we can say is that certain membranes are more permeable to water than to the substances in solution, and are moreover very unequally permeable to the various substances in solution. As a rule a membrane is much more permeable to a solute whose molecule is of small dimensions. Molecules of salt, for instance, pass through such a membrane much more quickly than do those of sugar. The term "osmotic membrane" should therefore in all cases replace that of "semi-permeable membrane."

Osmotic membranes behave exactly like colloids. The resistance which they oppose to the passage of different substances varies with the nature of the liquid or solute concerned. There is no real difference between the passage of a solution through an osmotic membrane and its diffusion through a colloid. The protoplasm of a living organism, being a colloid, acts exactly like an osmotic membrane so far as regards the distribution of solutions and substances in solution. {46}

The diffusion of molecules through a colloid, a plasma, or a membrane is governed by laws precisely analogous to Ohm's law, which governs the transport of electricity. The intensity or rapidity of diffusion is proportional to the difference of osmotic pressure, and varies inversely with the resistance.

In the case of molecular diffusion, however, the rapidity of diffusion depends also on the size and nature of the molecules of the diffusing substance. The theory of the resistance of the various plasmas and membranes to diffusion has been but little understood; we can discover hardly any reference to it in the literature of the subject.

The laws of diffusion apply equally to the diffusion of ions. Nernst has shown that there is a difference of electric potential at the surface of contact of two electrolytic solutions of different degrees of concentration. Both the positive and negative ions of the more concentrated solution pass into the less concentrated solution, but the ions of one sign will pass more rapidly than those of the other sign, because being smaller, they meet with less resistance.

The resistance of the medium plays a most important part in all the phenomena of diffusion. When two solutions of different concentration come into contact, the interchange of molecules and ions which occurs is unequal owing to the differences in resistance. Hence both solutions become modified not only in concentration but also in composition. It has long been known that diffusion can cause the decomposition of certain easily decomposed substances, and it would appear probable that diffusion is also capable of producing new chemical combinations.

The separation of the liberated ions in consequence of the unequal resistance which they meet with in the medium they traverse often determines chemical reaction. This ionic separation is a fertile agent of chemical transformation in the living organism, and may be the determinant cause in those chemical reactions which constitute the phenomena of nutrition.

When different liquids come into contact there are two distinct series of phenomena, those due to osmotic pressure and those due to differences of chemical composition. Even {47} with isotonic solutions there will be a transfer of the solutes if these are of different chemical constitution. Take, for instance, two isotonic solutions, one of salt and another of sugar. When these are brought into contact there is no transference of water from one solution to the other, but there is a transference of the solutes. In the salt solution the osmotic pressure of the sugar is zero. Hence the difference of osmotic pressure of the sugar in the two solutions will cause the molecules of sugar to diffuse into the salt solution. For the same reason the salt will diffuse into the sugar solution.

A disregard of this fact, that a solute will always pass from a solution where its osmotic pressure is high, into one where its osmotic pressure is low, is a frequent source of error. Thus it is said to be contrary to the laws of osmosis that solutes should pass from the blood, with its low osmotic pressure, into the urine, where the general osmotic pressure is higher; the more so because in consequence of the exchange the osmotic pressure of the urine is still further increased. Such an exchange, it is argued, is contrary to the ordinary laws of physics, and can therefore only be accomplished by some occult vital action. This, however, is not the fact, as is proved by experiment.

Consider an inextensible osmotic cell containing a solution of sugar, the walls of the cell being impermeable to sugar but permeable to salt. Let us plunge such a cell into a solution of salt, which has a lower osmotic pressure than the sugar solution. Since the walls of the cell are inextensible, the quantity of water in the cell cannot increase. The salt, however, will pass into the cell, since the osmotic pressure of the salt is greater on the outside than on the inside, and the walls are permeable to the molecules of salt. This passage will continue until the osmotic pressure of the salt is equal inside and outside the cell; at the same time the total osmotic pressure within the cell will have increased, in spite of its being originally greater than the osmotic pressure outside.

Plasmolysis.--We all know that a cut flower soon dries {48} up and fades. When, however, we place the shrivelled flower in water, the contracted protoplasm swells up again and refills the cells, which become turgid, and the flower revives. This phenomenon is due to the fact that vegetable protoplasm holds in solution substances like sugars and salts which have a high osmotic pressure. Consequently water has a tendency to penetrate the cellular walls of plants, to distend the cells and render them turgescent. De Vries has used this phenomenon for the measurement of osmotic tension. He employs for this purpose the turgid cells of the plant Tradescantia discolor. The cells are placed under the microscope and irrigated with a solution of nitrate of soda. On gradually increasing the concentration of the solution there comes a moment when the protoplasmic mass is seen to contract and to detach itself from the walls of the cell. This phenomenon, which is known as plasmolysis, occurs at the moment when the solution of nitrate of soda begins to abstract water from the protoplasmic juice, i.e. when the osmotic tension of the nitrate of soda becomes greater than that of the protoplasmic liquid. So long as the osmotic tension of the soda solution is less than that of the protoplasm, there will be a tendency for water to penetrate the cell wall and swell the protoplasm. When the osmotic tension of the solution which bathes the cell is identical with that of the cellular juice, there is no change in the volume of the protoplasm. In this way we are able to determine the osmotic pressure of any solution. We have only to dilute the solution till it has no effect on the protoplasm of the vegetable cells. Since the osmotic tension of this protoplasm is known, we can easily calculate the osmotic tension of the solution from the degree of dilution required.

Red Blood Corpuscles as Indicators of Isotony.--In 1886, Hamburger showed that the weakest solutions of various substances which would allow the deposition of the red blood cells, without being dilute enough to dissolve the hemoglobin, were isotonic to one another, and also to the blood serum, and to the contents of the blood corpuscles. This is Hamburger's method of determining the osmotic {49} tension of a liquid. The diluted solution is gradually increased in strength until, when a drop of blood is added to it, the corpuscles are just precipitated, and no hemoglobin is dissolved.

The Hematocrite.--In 1891, Hedin devised an instrument for determining the influence of different solutions on the red blood corpuscles. This instrument, the hematocrite, is a graduated pipette, designed to measure the volume of the globules separated by centrifugation from a given volume of blood under the influence of the liquid whose osmotic pressure is to be measured. The method depends on the principle that solutions isotonic to the blood corpuscles and to the blood serum will not alter the volume of the blood corpuscles, whereas hypertonic solutions decrease that volume.

Action of Solutions of Different Degrees of Concentration on Living Cells.-- We have just seen that a living cell, whether vegetable or animal, is not altered in volume when immersed in an isotonic solution that does not act upon it chemically. When immersed in a hypertonic solution, it retracts; in a slightly hypotonic solution it absorbs water and becomes turgescent, while in a very hypotonic solution it swells up and bursts. In a hypertonic solution the red blood cells retract and fall to the bottom of the glass, the rapidity with which they are deposited depending on the amount of retraction. In a hypotonic solution they swell up and burst, the hemoglobin dissolving in the liquid and colouring it red. This is the phenomenon of hematolysis. According to Hamburger, the serum of blood may be considerably diluted with water before producing hematolysis. Experimenting with the blood of the frog, he found that the globules remained intact in size and shape when irrigated with a salt solution containing .64 per cent. of salt, this solution being isotonic with the frog's blood serum. On the other hand, they did not begin to lose their hemoglobin till the proportion of salt was reduced to below .22 per cent. Thus frog's serum may be diluted with 200 per cent. of water before producing hemaatolysis. In mammals the blood corpuscles remain invariable in a salt solution of about .9 per cent., and begin to lose their {50} hemoglobin approximately in a .6 per cent. solution. A solution of .9 per cent. of NaCl is therefore isotonic to the contents of the red blood corpuscles, to the serum of the blood, and to the cells of the tissues. It by no means follows that the cells of the blood and tissues undergo no change when irrigated with a .9 per cent. solution of chloride of sodium. They do not lose or gain water, it is true, and they retain their volume and their specific gravity. But they do undergo a chemical alteration, by the exchange of their electrolytes with those of the solution. Hamburger has pointed out that in mammals the shape of the red corpuscles is altered in every liquid other than the blood serum; even in the lymph of the same animal there is a diminution of the long diameter, and an increase of the shorter diameter, while the concave discs become more spherical.

All the cells of a living organism are extremely sensitive to slight differences of osmotic pressure--the cells of epithelial tissue and of the nervous system as well as the blood cells. For instance, the introduction of too concentrated a saline solution into the nasal cavity will set up rhinitis and destroy the terminations of the olfactory nerves. Pure water, on the other hand, is itself a caustic. There is a spring at Gastein, in the Tyrol, which is called the poison spring, the "Gift-Brunnen." The water of this spring is almost absolutely pure, hence it has a tendency to distend and burst the epithelium cells of the digestive tract, and thus gives rise to the deleterious effects which have given it its name. Ordinary drinking water is never pure, it contains in solution salts from the soil and gases from the atmosphere. These give it an osmotic pressure which prevents the deleterious effects of a strongly hypotonic liquid. During a surgical operation it is of the first importance not to injure the living surfaces by flooding them with strongly hypertonic or hypotonic solutions. This precaution becomes still more important when foreign liquids are brought into contact with the delicate cells of the large surfaces of the serous membranes. Gardeners are well aware of the noxious influence of a low osmotic pressure. They water the soil around the roots of a plant, so that the water may take up {51} some of the salts from the soil before being absorbed by the plant. Pure water poured over the heart of a delicate plant may burst its cells owing to its low osmotic pressure. In many medical and surgical applications, on the other hand, a low osmotic pressure is of advantage. Thus, in order to remove the dry crusts of eczema and impetigo, the most efficacious application is a compress of cotton wool soaked in warm distilled water. Under the influence of such a hypotonic solution the dry cells rapidly swell up, burst, and are dissolved.

Cooking is also very much a question of osmotic pressure. If salt is put into the water in which potatoes and other vegetables are boiled, osmosis is set up and a current of water passes from the vegetable cells to the salt water. The cellular tissue of the vegetable becomes contracted and dried, and the membranes become adherent, the vegetable loses weight and becomes difficult of digestion, in consequence of its hard and waxy consistency, which prevents the action of the digestive juices. Vegetables should be cooked in soft water, and should be salted after cooking. When so treated, a potato absorbs water, the cells swell up, the skin bursts, the grains of starch also swell up and burst, and the pulp becomes more friable. The digestive juice is thus able to penetrate the different parts of the vegetable rapidly, and digestion is facilitated. Any one can easily prove for himself that a potato boiled in salt water diminishes in weight, whilst its weight increases when it is cooked in soft water.

The method of cryoscopy is also of considerable service in forensic medicine. As shown by Carrara, the cryoscopy of the blood is an important aid in determining the question whether a body found in the water was thrown in before or after death. In the former case the concentration of the blood will be much diminished. In certain experiments on dogs the cryoscopic examination of the blood showed a freezing point of -.6?C. The dog was then drowned, when the freezing point of the blood in the left ventricle was increased to -.29?C., and that in the right ventricle to -.42?C. On the other hand, when a dog was killed before being thrown into the water, the {52} osmotic pressure of the blood was hardly decreased even after an immersion of 72 hours. In the case of persons or animals drowned in sea water, a similar alteration of the point of congelation is observed, but in the reverse direction. In this case the osmotic pressure is raised considerably in those who are drowned, whereas no such rise is observed in those who are thrown into the sea after death.

The circulation of the sap in plants and trees is also in great part due to osmotic pressure. The aspiration of the water from the soil is due to the intracellular osmotic pressure in the roots, which causes the sap to rise in the stem of a plant as it would in the tube of a manometer. From a knowledge of the osmotic pressure of the intracellular liquid of the roots, we may calculate the height to which the sap can be raised in the trunk of a tree, i.e. the maximum height to which the tree can possibly grow. Suppose, for instance, the plasma of the rootlets has an osmotic pressure of six atmospheres, corresponding to that of a 9 per cent. solution of sugar. A pressure of six atmospheres is equal to the weight of a column of water 6 ?.76 ?13.596 = 61.95 metres high. This, then, is the maximum height to which this osmotic pressure is able to lift the sap. That is to say, a tree whose rootlets contain a solution of sugar of 9 per cent. concentration, or its equivalent, can grow to a height of 62 metres.

Cryoscopy is also of great use in practical medicine, more especially for the examination of the urine. The freezing point of urine varies from -1.26?C. to -2.35? Koryani has studied the ratio of the point of congelation of urine to that of a solution containing an equal quantity of chloride of sodium. He finds that the ratio (freezing point of urine) / (freezing point of NaCl) increases when the circulation through the tubules of the kidney is diminished.

Hans Koeppe has shown that the hydrochloric acid of the gastric juice is produced by the osmotic exchanges between the blood and the gastric contents. The ion Na^+ of the salt in the stomach contents exchanges with an ion H^+ of the monobasic salts of the blood, $NaHCO_3 + NaCl = HCl + Na_2CO_3$. {53}

Influence of Muscular Contraction on the Intramuscular Osmotic Pressure.--
When a muscle is immersed in an isotonic salt solution it does not change in weight. In a hypertonic solution it loses weight in consequence of a loss of water, which passes from the muscle into the solution to equalize the osmotic pressure. It gains weight in a hypotonic solution, the water current setting towards the point of higher concentration. It is easy, therefore, to tell whether the osmotic pressure in a muscle is above or below that of a given solution, by observing whether the muscle gains or loses weight when immersed in it. Thus we may measure the osmotic pressure in a muscle by finding a salt solution in which the muscle neither gains nor loses weight. In this way we have been able to prove that the osmotic pressure of a tired muscle is higher than that of the normal muscle. Our experiments were carried out on the muscles of frogs. After having pithed the frog, one of the hind legs is removed by a single stroke of the scissors. The leg is skinned, dried with blotting paper, and weighed. It is then placed in a salt solution whose freezing point is -.53?C. At 15?C. such a solution has an osmotic pressure of 6.6 atmospheres. We next proceed to determine the osmotic pressure of the corresponding leg after it has been tired by muscular work. For this it is stimulated by an intermittent faradic current passing once a

second for five minutes. The leg is then skinned, dried, weighed, and placed in the same salt solution. After eight hours' immersion the legs are weighed again. The following are the results of six experiments, the numbers representing fractions of the original weight:--

Change of weight of untired leg--

After 8 hours -.000. After 16 hours -.000. After 24 hours -.006.

Change of weight of stimulated leg--

After 8 hours +.050. After 16 hours +.080. After 24 hours +.101.

{54}

This result shows that muscular work provoked by electric stimulation noticeably increases the osmotic pressure of the muscle.

In order to discover the exact osmotic pressure in the stimulated muscles we repeated the series of experiments, using more and more concentrated solutions. In a solution whose freezing point was -.57? we obtained the following values:--

Change of weight of untired leg--

After 8 hours -.000. After 16 hours -.004. After 24 hours -.006.

Change of weight of stimulated leg--

After 8 hours +.039. After 16 hours +.072. After 24 hours +.099.

Finally, in a solution freezing at -.72? i.e. with an osmotic pressure at 15?C. of 9.176 atmospheres, we obtained the following mean values for the untired leg:--

After 8 hours -.04. After 16 hours -.05. After 24 hours -.05.

In this solution, freezing at -.72?C., some of the stimulated muscles showed no diminution in weight, while others showed a very small diminution, and others again a slight augmentation, the maximum increase being .085 of the initial weight. The solution is therefore practically isotonic with the stimulated muscle.

In this case the elevation of the intramuscular osmotic pressure produced by the electrical excitation and the muscular contractions was therefore 2.5 atmospheres, or more than 2.6 kilogrammes per square centimetre of surface.

I made further experiments in order to discover whether the variation in osmotic pressure depended on the duration of {55} the muscular contraction. For this purpose I used a solution freezing at -.53?C. and immersed in it untired muscles, and muscles which had been electrically excited for two, four, and six minutes respectively. The following are the results:--

Untired muscles. Muscles stimulated once a second during 2 Minutes. 4 Minutes. 6 Minutes. .000 +.026 +.084 +.094 +.001 +.034 +.065 +.093 +.005 +.045 +.079 +.097 .000 +.037 +.070 +.095 .000 +.032 +.072 +.096

Mean of all the observations--

+.0012 +.0348 +.074 +.095

These experiments show clearly that the osmotic intramuscular pressure rises in proportion to the duration of the electrical stimulation.

In order to determine the influence of the work accomplished by the muscle on the elevation of the osmotic pressure, I made the following experiment. The two hind legs of a frog were submitted to the same electrical excitation, one leg being left at liberty, and the other being stretched by a hundred-gramme weight, acting by a cord and pulley. After exciting them electrically for five minutes, the legs were immersed for twenty-four hours in a saline solution freezing at .53?C. The free limb showed an augmentation of .085 of the initial weight, and the stretched limb an increase of .106 of the initial weight. It is evident, therefore, that the osmotic pressure increases with the amount of work done by a muscle.

Briefly, then, the results of our experiments are as follow:--

1. Muscular contraction electrically produced causes an increase of the osmotic pressure in a muscle.

2. The intramuscular osmotic pressure may reach, or even exceed, 2.5 atmospheres, or 2.6 kilogrammes per square centimetre of surface.

3. When a muscle is made to contract once a second, the {56} elevation of the osmotic pressure increases with the number of contractions.

4. The intramuscular osmotic pressure increases with the work done by the muscle.

5. Fatigue is caused by the increase of osmotic pressure in a contracting muscle.

The Field of Diffusion.--Just as Faraday introduced the conception of a field of magnetic force and a field of electric force to explain magnetic and electrical phenomena, so we may elucidate the phenomena of diffusion by the conception of a field of diffusion, with centres or poles of diffusive force. If we consider a solution as a field of diffusion, any point where the concentration is greater than that of the rest may be considered as a centre of force, attractive for the molecules of water, and repulsive for the molecules of the solute. In the same way any point of less concentration may be regarded as a centre of attraction for the molecules of the solute, and a centre of repulsion for the molecules of water.

A field of diffusion may be monopolar or bipolar. A bipolar field has a hypertonic pole or centre of concentration, and a hypotonic pole or centre of dilution. By analogy with the magnetic and electric fields we may designate the hypertonic pole as the positive pole of diffusion, and the hypotonic as the negative pole. {57}

The positive and negative poles and the lines of force in the field of diffusion may be illustrated by the following experiment. A thin layer of salt water is spread over an absolutely horizontal plate of glass. If now we take a drop of blood, or of Indian ink, and drop it carefully into the middle of the salt solution, we shall find that the coloured particles will travel along the lines of diffusive force, and thus map out for us a monopolar field of diffusion, as in Fig. 3 a. Again, if we place two similar drops side by side in a salt solution, their lines of diffusion will repel one another, as in Fig. 4.

Now let us put into the solution, side by side, one drop of less concentration and another of greater concentration than the solution. The lines of diffusion will pass from one drop to the other, diverging from the centre of one drop and converging towards the centre of the other (Fig. 3 b). In this manner we are able to obtain diffusion fields analogous to the magnetic fields between poles of the same sign and poles of opposite signs.

The conception of poles of diffusion is of the greatest importance in biology, throwing a flood of light on a number of phenomena, such as karyokinesis, which have hitherto been regarded as of a mysterious nature. It also enables us to appreciate the role played by diffusion in many other biological phenomena. Consider, for example, a centre of anabolism in a living organism. Here the molecules of the living protoplasm are in process of construction, simpler molecules being united and built up to form larger and more complex groups. As a result of this aggregation the number of molecules in a given area is diminished, i.e. the concentration and the osmotic pressure fall, producing a hypotonic centre of diffusion. We may thus regard every centre of anabolism as a negative pole of diffusion. {58}

Consider, on the other hand, a centre of catabolism, where the molecules are being broken up into fragments or smaller groups. The concentration of the solution is increased, the osmotic pressure is raised, and we have a hypertonic centre of diffusion. Every centre of catabolism is therefore a positive pole of diffusion. Similar considerations as to the formation and breaking up of the molecules in anabolism and catabolism apply to polymerization.

The diffusion field has similar properties to the magnetic and the electric field. Thus there is repulsion between poles of similar sign, and attraction between poles of different signs. A simple experiment will show this. A field of diffusion is made by pouring on a horizontal glass plate a 10 per cent. solution of gelatine to which 5 per cent. of salt has been added. The gelatine being set, we place side by side on its surface two drops, one of water, and one of a salt solution of greater concentration than 5 per cent. We have thus two poles of diffusion of contrary signs, a hypotonic pole at the water drop, and a hypertonic pole at the salt drop. Diffusion immediately begins to take place through the gelatine, the drops become elongated, advance towards one another, touch, and unite. If, on the contrary, the two neighbouring drops are both more concentrated or both less concentrated than the medium, they exhibit signs of repulsion as in Fig. 4.

Diffusion not only sets up currents in the water and in the solutes, but it also determines movements in any particles that may be in suspension, such as blood corpuscles, particles of Indian ink, and the like. These particles are drawn along with the water stream which passes from the hypotonic centres or regions toward those which are hypertonic.

These considerations suggest a vast field of inquiry in biology, pathology, and therapeutics. Inflammation, for example, is characterized by tumefaction, turgescence of the tissues, and redness. The essence of inflammation would appear to be destructive dis-assimilation with intense catabolism. We have seen that a centre of catabolism is a hypertonic focus of diffusion. Hence the osmotic pressure in an inflamed region is increased, turgescence is produced, and {59} the current of water carries with it the blood globules which produce the redness.

The phenomenon of agglutination may also possibly be due to osmotic pressure, a positive centre of diffusion attracting and agglomerating the particles held in suspension.

Tactism and Tropism.--The phenomena of tactism and tropism may also be partly explained by the action of these diffusion currents of particles in suspension, these polar attractions and repulsions. In all experiments on this subject we should take into account the possible influence of osmotic pressure, since many of the causes of tactism or tropism also modify the osmotic pressure at the point of action, and it is possible that this modification is the true cause of the phenomenon. Osmotactism and osmotropism have not as yet been sufficiently studied.

The six negative poles of diffusion are coloured with Indian ink. The positive pole in the centre is uncoloured and is formed by a drop of KNO3 solution.]

Thus it may be said that osmotic pressure dominates all the kinetic and dynamic phenomena of life, all those at least which are not purely mechanical, like the movements of respiration and circulation. The study of these vital phenomena is greatly facilitated by the conception of the field of diffusion and poles of diffusion, and of the lines of force, which are the trajectories of the molecules of the solutes, and the particles and globules in suspension.

The Morphogenic Effects of Diffusion.--Many interesting experiments may be made showing variations of the lines of force in a field of diffusion, and how liquids subjected only to differences of osmotic pressure diffuse and mix with one {60} another in definite patterns. When a liquid diffuses in another undisturbed by the influence of gravity, it produces figures of geometric regularity, and we may thus obtain figures and forms of infinite variety. The following is our method of procedure. A glass plate is placed absolutely horizontal and is covered with a thin layer of water or of saline solution. Then with a pipette we introduce into the solution, in a regular pattern, a number of drops of liquid coloured with Indian ink. A wonderful variety of patterns and figures may be obtained by employing solutions of different concentration and varying the position of the drops.

Instead of the water or salt solution, we may spread on the plate a 5 or 10 per cent. solution of gelatine, containing various salts in solution. If now we sow on this gelatine drops of various solutions which give colorations with the salts in the gelatine, we may obtain forms of perfect regularity, presenting most beautiful colours and contrasts. The drops, of course, must be placed in a symmetrical pattern. In this way we may obtain an endless number of ornamental figures.

In order to cover a lantern slide 8?cm. ?10 cm., about 5 c.c. of gelatine is required. To this amount of gelatine we add a single drop of a saturated solution of salicylate of sodium, and spread the liquid gelatine evenly over the plate. When the gelatine has set, we put the plate over a diagram, a hexagon for instance, and place a drop of ferrous sulphate solution at each of the six angles. The drops immediately diffuse {61} through the gelatine, and the result after a time is the production of a beautiful purple rosette. The gelatine must be carefully covered to prevent its drying until the diffusion is complete. The preparation may then be dried and mounted as a lantern slide, and will give the most brilliant effect on projection. If the gelatine has been treated with a drop of potassium ferrocyanide solution instead of salicylate of sodium, a few drops of FeSO4 will give a blue pattern. Or we may treat the gelatine with ferrocyanide of potassium and salicylate of sodium mixed, and thus obtain an intermediary colour on the addition of FeSO4. We may, indeed, vary indefinitely the nature and concentration of the solution, as well as the number and position of the drops. The results have all the charm of the unexpected, which adds greatly to the interest of the experiment.

These experiments are not merely a scientific toy. They show us the possibility, hitherto unsuspected, that a vast number of the forms and colours of nature may be the result of diffusion. Thus many of the phenomena of life, hitherto so mysterious, present themselves to us as merely the consequences of the diffusion of one liquid into another. One cannot help hoping that the study of diffusion will throw still further light on the subject.

If a number of spheres, each capable of expansion and deformation, are produced simultaneously in a liquid, they will form polyhedra when they expand by growth. This is the {62} precise architecture of a vast number of living organisms and tissues, which are formed by the union of microscopic polyhedra or cells. A section of such a polyhedral structure would appear as a tissue of polygons. It is interesting to note that the simple process of diffusion will produce such structures under conditions closely allied to those which govern the development of the tissues of a living organism.

We may obtain this cellular structure by a simple experiment. On a glass plate we spread a 5 per cent. solution of pure gelatine, and when set sow on it a number of drops of a 5 to 10 per cent. solution of ferrocyanide of potassium. The drops must be placed at regular intervals of 5 mm. all over the plate. When these have been allowed to diffuse and the gelatine has dried, we obtain a preparation which exactly resembles the section of a vegetable cellular tissue (Fig. 9). The drops have by mutual pressure formed polygons, which appear in section as cells, with a membranous envelope, a {63} nucleus, and a cytoplasm, which is in many cases entirely separated from the membrane. These cells when united form a veritable tissue, in all respects similar to the cellular structure of a living organism.

[Illustration: FIG. 9.--Tissue of artificial cells formed by the diffusion in gelatine of drops of potassium ferrocyanide.]

In the preparation showing artificial cells the cellular structure is not directly visible until the gelatine has dried. One sees only a gelatinous mass analogous to the protoplasm of a living organism. This mass is nevertheless organized, or at least in process of organization, as we may see by the refraction when its image is projected on the screen.

During the cell-formation, and as long as there is any difference of concentration in the gelatine, each cell is the arena of active molecular movement. There is a double current, as in the living cell, a stream of water from the periphery to the centre, and of the solute from the centre to the periphery. This molecular activity--the life of the artificial cell--may be prolonged by appropriate nourishment, {64} i.e. by continually repairing the loss of concentration at the centre of the cell.

The life of the artificial cell may also be prolonged by maintaining around it an appropriate medium. If we prematurely dry such a preparation of artificial cells, the molecular currents will cease, to recur again when we restore the necessary humidity to the preparation. This to my mind gives us a most vivid picture of the conditions of latent life in seeds and many rotifera.

These artificial cells, like living organisms, have an evolutionary existence. The first stage corresponds to the process of organization, the gelatine representing the blastema, and the drop the nucleus. Thus the cell becomes organized, forming its own cytoplasm and its own enveloping membrane.

The second stage in the life of this artificial cell is the period during which the metabolism of the cell is active and tends to equalize the concentration of the liquid in the cell and in the surrounding medium.

The third stage is the period of decline. The double molecular current gradually slows down as the difference of concentration decreases between the cell contents and its entourage. When this equality of concentration has become complete the molecular currents cease, the cell has terminated its existence; it is dead. The currents of substance and of energy have ceased to flow--the form only remains.

These artificial cells are sensible to most of the influences which affect living organisms. Like living cells they are influenced both in their organization and in their development by humidity, dryness, acidity, or alkalinity. They are also greatly affected by the addition of minute quantities of chemical substances either to the gelatinous blastema or to the drops which represent the primary nuclei. We may in this way obtain endless varieties, nuclei which are opaque or transparent, with or without a nucleolus, and cells containing homogeneous cytoplasm without a nucleus. We may also obtain cells with cytoplasm filling the whole of the cellular cavity or separated from the cell-membrane. We may obtain {65} cells imitating all the natural tissues, cells without a membranous envelope, cells with thick walls adhering to one another, or cells with wide intracellular spaces.

The forms of these artificial cells depend on the number and relative position of the drops which represent the nuclei, and on the molecular concentration or osmotic tension of the solution. The number of the cellular polyhedra is determined by the number of centres of diffusion. The magnitude of the dihedral angles, from which radiate three and occasionally four walls, depends on the position of the hypertonic poles of diffusion. The curvature of a surface is determined by the differences of concentration on either side. Between isotonic solutions the surface is plane, whilst it is curved between solutions of different osmotic pressures, the convexity being directed towards the hypertonic solution.

[Illustration: FIG. 11.--Liquid cells with a fringe of cilia, obtained by sowing coloured drops of concentrated salt solution in a weaker salt solution. The contents of the cells have undergone segmentation.]

The time required for these artificial cells to grow varies from two to twenty-four hours, according to the concentration of the gelatine, the growth being most rapid in dilute solutions.

Similar cells may be produced in water. If we pour a thin layer of water on a horizontal plate, and with a pipette {66} sow in it a number of drops of salt water coloured with Indian ink, we may obtain artificial cells composed entirely of liquid, having the same characters as those produced in a gelatinous solution.

It is possible by liquid diffusion to produce not only ordinary cells but ciliated cells. If we spread a layer of salt water on a horizontal glass plate, and sow in it drops of Indian ink, artificial cells are produced by diffusion. At the edge of the preparation there is often to be seen a sort of fringe, analogous to the cilia of living cells (Fig. 11).

These tissues of artificial cells demonstrate the fact that inorganic matter is able to organize itself into forms and structures analogous to those of living organisms under the action of the simple physical forces of osmotic pressure and diffusion. The structures thus produced have functions which are also analogous to those of living beings, a double current of diffusion, an evolutionary existence, and a latent vitality when desiccated or congealed.

* * * * *

{67}

CHAPTER VI

PERIODICITY

Periodic Precipitation.--A phenomenon is said to be periodic when it varies in time and space and is identically reproduced at equal intervals. We are surrounded on all sides by periodic phenomena; summer and winter, day and night, sleep and waking, rhythm and rhyme, flux and reflux, the movements of respiration and the beating of the heart, all are periodic. Our first sorrows were appeased by the periodic rhythm of the cradle, and in our later years the periodic swing of the rocking-chair and the hammock still soothe the infirmities of old age.

Sound is a periodic movement of the atmosphere which brings to us harmony and melody. Light consists of periodic undulations of the ether which convey to us the beauty of form and colour. Periodic ethereal waves waft to us the wireless message through terrestrial space and the radiant energy of the sun and stars.

It is therefore not to be wondered at that the phenomena of diffusion are also periodic. According to Professor Quinke of Heidelberg, the first mention of the periodic formation of chemical precipitates must be attributed to

Runge in 1885. Since that time these precipitates have been studied by a number of authors, and particularly by R. Liesegang of Dusseldorf, who in 1907 published a work on the subject, entitled On Stratification by Diffusion.

In 1901 I presented to the Congress of Ajaccio a number of preparations showing concentric rings, alternately transparent and opaque, obtained by diffusing a drop of potassium ferrocyanide solution in gelatine containing a trace of ferric {68} sulphate. At the Congress of Rheims in 1907 I exhibited the result of some further experiments on the same subject.

These periodic precipitates may be obtained from a great number of different chemical substances. The following is the best method of demonstrating the phenomenon. A glass lantern slide is carefully cleaned and placed absolutely level. We then take 5 c.c. of a 10 per cent. solution of gelatine and add to it one drop of a concentrated solution of sodium arsenate. This is poured over the glass plate whilst hot, and as soon as it is quite set, but before it can dry, we allow a drop of silver nitrate solution containing a trace of nitric acid to fall on it from a pipette. The drop slowly spreads in the gelatine, and we thus obtain magnificent rings of periodic precipitates of arsenate of silver, with which any one may easily repeat the experiments detailed in this chapter.

Circular Waves of Precipitation.--The wave-front of the periodic rings of precipitates is always perpendicular to the rays of diffusion. The distance between the rings depends on the concentration of the diffusing solution. The greater the fall of concentration, the less is the interval between the rings. Each ring represents an equipotential line in the field of diffusion. These equipotential lines of diffusion give us the best and most concrete reproduction of the mode of propagation of periodic waves in space. They are, in fact, a visible diagram of the propagation of the waves of light and sound. Occasionally we may observe in the gelatine the simultaneous propagation of undulations of different wave-length, just as we have them in the ether and the air. These diffusion wavelets {69} give us a very beautiful representation of the simultaneous propagation of undulations of different wave-length in the same medium.

Like waves of light and sound, these waves of diffusion are refracted when they pass from one medium into another of a different density, where they have a different velocity. When, for instance, a diffusion wave passes from a 5 per cent. solution of gelatine into a 10 per cent. solution, the wave-front is retarded, the retardation being proportional to the length of the path through the denser medium. Hence the wave-front is flattened, the curvature of the refracted wave being less than that of the original wave of diffusion. The contrary is the case when the wave-front passes into a medium where its velocity is greater. The middle of the wave-front now travels faster than the flanks, and the curvature is increased.

These diffusion rings furnish us with most excellent diagrams of refraction at a "diopter," i.e. a spherical surface separating two media of different densities. Fig. 14 shows the refraction at a convergent diopter, i.e. a surface where the denser medium is convex. The diffusion waves in this case emanate from the principal focus of the diopter, and therefore become plane on passing through the convex surface of the denser gelatine.

These periodic diffusion rings also illustrate the phenomena of colour diffraction. Diffusion waves of different {70} wavelength are unequally refracted by a gelatine lens. Hence rings of different wave-length which, originating at the same spot, are at first concentric, are no longer parallel after passing through a gelatine lens. A convergent lens which will change the long spherical incident waves into shorter plane waves, will transform the short incident waves into concave waves whose curvature is opposite to that of the original waves, i.e. it will transform a divergent into a convergent beam. This is an illustration of what is called the aberration of refrangibility.

In the same way we may demonstrate the course of diffusion waves through a gelatine prism, showing the refraction on their incidence and again on emergence. The prism is made of a stronger gelatine solution, which is more refractive than the gelatine around it. The waves of diffusion whilst traversing the prism are retarded, and this retardation is greatest at the base where the passage is longer. Hence the wave-front is tilted towards the base of the prism, and this tilting is repeated when the wave-front leaves the prism.

If we examine diffusion waves of different wave-length on their emergence from the gelatine prism, we shall see that they cut one another. With a dense prism, the wave-front of the shorter waves is more tilted towards the base than the wave-front of the longer waves. For diffusion as for light the shorter waves are the most refracted. Both refraction and dispersion are due to the unequal resistances of the medium to undulatory movements of different periodicity.

Diffraction.--When light traverses a minute orifice, instead {71} of passing on in a straight line, it spreads out like a fan, forming a diverging cone of light, just as if the orifice were itself a luminous point. This is the phenomenon of diffraction which has hitherto been considered incompatible with the emission theory of light. Diffusion waves may also be made to pass through a narrow orifice, when they will behave exactly like the waves of light. The new waves radiate from the orifice like a fan, instead of giving a cone of waves bounded by lines passing through the circumference of the orifice and the original centre of radiation. Thus on passing through a small orifice diffusion waves exhibit the phenomenon of diffraction just as light waves do.

Interference.--The phenomenon of interference may also be illustrated by waves of diffusion. If on a gelatine plate we produce two series of diffusion waves from two separate centres, we get at certain points an appearance corresponding to the interference of two sets of light waves. This appearance is best shown by sowing on the gelatine film a straight row of drops equidistant from one another. It should be remarked that this phenomenon of the production of circles of precipitate separated by transparent spaces, although periodic, is not of necessity vibratory or undulatory. It would thus appear that periodic phenomena may be propagated through space without vibratory or oscillatory motion. If we submit to a critical examination the various experiments which have established the undulatory theory of light, we find that they do indeed demonstrate the periodic nature of light, but in no wise prove that light is a vibratory movement of the ether. {72} On the contrary, the hypothesis that light is propagated by vibratory movements is open to many objections. Even the Zeeman effect, although it may tend to establish the fact that light is produced by vibratory movement, by no means proves that it is propagated in the same manner. When the theory was accepted that the transmission of light was periodic it was supposed that this periodic transmission could only be vibratory or undulatory in character, since waves or vibrations were the only periodic phenomena known at that time. We now know that there are other means of periodic transmission which are apparently not undulatory. The periodic precipitates produced by diffusion show us the transmission of spherical waves through space, which follow the laws of light, although the periodic phenomenon is apparently emissive rather than vibratory.

It will be remembered that Newton considered light to be produced by projectile-like particles emanating from a centre, and proceeding in straight lines in all directions. This emission theory of light was abandoned in favour of Huygens' undulatory theory.

It was said that the phenomena of interference and diffraction could not be explained by the theory of emission, while the undulatory theory gave a simple explanation. The scientific mind was unable to conceive the idea of emission and periodicity as taking part in the same phenomenon. The savants and thinkers who have meditated on this question have always considered the theory of emission and that of periodicity as incompatible. Nevertheless, we are here in presence of a phenomenon in which emission and periodicity exist simultaneously. The molecules emanating from our drop are diffused in straight radiating lines, and yet produce periodic precipitates which are subject to interference and diffraction like the undulations of Huygens.

The phenomena associated with the pressure of light, the {73} discovery of the cathode rays and the radiations of radium, together with the introduction of the electron theory of electricity, all seem to have brought again into greater prominence Newton's original conception of the emissionary nature of light.

Some of the phenomena of radiation can be explained only by the emission theory, and others by the undulatory theory of light. All these difficulties would be solved if we admitted the hypothesis that radiating bodies project electrons, which produce in the ether periodic waves similar to those formed in our gelatine films by the molecules of diffusion.

These diffusion films are of the greatest possible service in the practical teaching of optics. They place before the eye of the student a working model as it were of the undulations of light. When projected on the screen, they give excellent pictures of the phenomena of refraction, diffraction, and interference, and the simultaneous propagation of undulation of different wave-lengths, and they show in a visible manner the changes of wave-length in media of different densities.

Diffusion waves differ greatly in length, varying from several millimetres to 2 [mu]. Many are even shorter than this, too short to be separately distinguished even under the highest power of the microscope, when they give the effect of moir?or mother-of-pearl.

It is easy to construct a spectroscopic grating in this way with fine lines whose distance apart is of the order of a micron, separated by clear spaces. Every physical laboratory may thus produce its own spectroscopic gratings, rectilinear, circular, or of any desired form.

The most beautiful colour effects may be produced with these diffusion gratings, as we have shown at the Congress of Rheims in 1907. We have a considerable collection of these diffusion gratings, some with very fine lines, giving a very extended spectrum, and others with coarser striations which give a large number of small spectra.

This study of periodic precipitates is of the highest interest when we come to investigate the production of colour in natural objects, such as the wings of insects or the plumage of {74} birds. Many tissues have this lined or striated structure and exhibit interference colours like those of the periodic precipitates, their structure showing alternate transparent and opaque lines, whose width is of the order of a micron. This is the structure of muscle, and to this striated surface is also attributable many of the most beautiful colours of nature, the gleam of tendon and aponeurosis, the fire of scarab and beetle, the colours of the peacock, and the iridescence of the mollusc and the pearl. The study of liquid diffusion has given us an idea of the physical mechanism by which these striated tissues are produced, a mechanism which up to the present time has not been even suspected. Our experiments show how readily such striped or ruled structures may be produced in a colloidal solution by the simple diffusion of salts such as are found in every living organism.

To make a spectroscopic grating by diffusion we proceed as follows. We take 5 c.c. of a 10 per cent. solution of gelatine, and add to it one drop of a concentrated solution {75} of calcium nitrate. We spread the gelatine evenly over a plain glass lantern slide and allow it to set. After it is set, but before it dries, we place in the centre of the slide a drop of concentrated solution containing two parts of sodium carbonate (Na2CO3) to one of dibasic sodium phosphate (Na2HPO4). Tribasic sodium phosphate alone without the addition of the carbonate will also give good results. If the phosphate solution is placed on the gelatine in the form of a drop, we obtain circular periodic precipitates. If it is desired to make a rectilineal grating, we deposit the phosphate solution on the gelatine in a straight line by means of two parallel glass plates. In this way we may obtain lines of periodic precipitation to the number of 500 to 1000 per millimetre, forming gratings which produce most beautiful spectra.

Pearls and mother-of-pearl both owe their iridescence to a similar ruled structure, which is developed in the living tissue of a mollusc. They are, in fact, periodic precipitates of phosphate and carbonate of lime deposited in the colloidal organic substance of the mollusc. They have the same structure and the same chemical composition; they have the same physical properties, the glow, the fire, and the brilliancy of our spectroscopic gratings. In these experiments, indeed, we have realized the synthesis of the pearl, not only a chemical synthesis, but the synthesis of its structure and organism.

We have been able to make these periodic precipitates by the reaction of a great number of chemical substances, giving a bewildering variety of form and structure. Some of these recall the form of various organisms, and especially of insects, as may be seen in Fig. 18.

All the phenomena of life are periodic. The movement of heart and lungs, sleep and waking, all nervous phenomena, have a regular periodicity. It is possible that the study of these purely physical phenomena of periodic precipitation may give us the key to the causation of rhythm and periodicity in living beings.

Besides this periodic precipitation there appear to be other chemical reactions which are periodic. Professor Bredig of Heidelberg has lately described a curious phenomenon, the {76} periodic catalysis of peroxide of hydrogen by mercury. He thus describes his experiment: "We place in a perfectly clean test tube a few cubic centimetres of perfectly pure mercury. Upon this we pour 10 c.c. of a 10 per cent. solution of hydrogen peroxide. The mercury speedily becomes covered with a thin, brilliant bronze-coloured pellicle which reflects light. Then little by little catalysis of the hydrogen peroxide begins, with liberation of oxygen. After some time, from five to twenty minutes, the liberation of gas at the surface of the mercury ceases, the cloud formed by the gas bubbles disappears, and the bronze mirror at the surface of the mercury lights up with the glint of silver. There is a pause of one or more seconds, and then the catalytic action begins afresh, commencing at the edges of the mirror. The cloud is again formed and again disappears. This beautiful and surprising rhythmic phenomenon may continue at regular intervals for an hour or more."

A slight alkalinity of the liquid is necessary to start the phenomenon. This explains the retardation at the beginning {77} of the experiment, since the rhythmic catalysis cannot begin until the hydrogen peroxide has dissolved a little of the glass so as to render it slightly alkaline. The catalytic process may, however, be set going at once by adding a trace of potassium acetate to the solution.

We may even obtain a curve giving an automatic record of the periodicity of this catalytic action. For this purpose the oxygen given off is led to a manometer, which registers on a revolving drum the periodic variation in pressure. The curve thus obtained presents a remarkable resemblance to a tracing of the pulse. The frequency and character of the undulatory curve is modified by physical and chemical influences. Like circulation or respiration, periodic catalysis has its poisons, and exhibits signs of fatigue, and of paralysis by cold.

The rhythmic catalysis of Bredig produces an electrical current of action between the mercury and the water just like that produced by the rhythmic contraction of the heart, and this current may be registered in a similar way by means of the Einthoven galvanometer. Thus the heart-beat may be but an instance of rhythmic catalysis, since both produce the same phenomena, movement, chemical action, and periodic currents. In the chapter on physiogenesis we shall return to the study of this question and consider another rhythmic phenomenon which is the result of osmotic growth.

* * * * *

CHAPTER VII

COHESION AND CRYSTALLIZATION

Chemical affinity is the force which holds together the different atoms in a molecule. Cohesion is the force which holds together molecules which are chemically similar. Although physical science distinguishes three states of matter, solid, liquid, and gaseous, yet here as elsewhere there are no sharp dividing lines, but rather an absolute continuity. We have in fact many intermediate states; between liquids and gases there are the various conditions of vapour, and between liquids and solids we get viscous, gelatinous, and paste-like conditions. The only real difference between solids, liquids, and gases is the intensity of the force of cohesion, which is considerable in solids, feeble in liquids, and absent in gases.

A living organism is the arena in which are brought into play the opposing forces of cohesion and disintegration. The study of cohesion is therefore a vital one for the biologist, and especially cohesion under the conditions which obtain in living beings, viz. in liquids of heterogeneous constitution. The forces of cohesion brought into play under these conditions may be beautifully illustrated by a simple experiment. We take a plate of glass, well cleaned and absolutely horizontal. On it we pour a layer of salt water, and in the middle we carefully drop a spot of Indian ink. The drop at once begins to diffuse, and we obtain a circular figure, like the monopolar field of diffusion already described, the rays of diffusion radiating from the centre in all directions.

If we keep the plate carefully protected from all disturbing influences, after some ten to twenty minutes we shall see the coloured particles returning on their path, and the centre of {79} the drop becoming more and more black. Each line of force becomes segmented into granules, which gradually increase in size, and approach nearer to one another and to the centre of the drop, until it assumes the mulberry appearance shown in the photograph (Fig. 19).

If we sow a number of drops of Indian ink in regular order on the surface of a salt solution, we obtain most beautiful patterns formed by the mutual repulsion of the drops. Figs. 20, 21, and 22 represent the successive aspects of seven drops of Indian ink thus sown on a layer of salt solution, and kept undisturbed long enough to allow of their evolution. Fig. 20 shows the aspect after two minutes, when the diffusion is almost complete. In Fig. 21, photographed after fifteen {80} minutes, the colouring matter has almost entirely reunited to form separate granulations; whilst in Fig. 22, taken after thirty minutes, these granulations are rearranged to form an agglomeration around the centre of each drop.

The following experiment, which is more difficult, will show the cohesive attraction of one drop for another. A plate of glass is adjusted absolutely horizontal, and covered as before with a layer of salt solution. On this we sow a number of drops of the same salt solution coloured with Indian ink. The drops must be of exactly the same concentration as the salt medium, so as to avoid any difference of osmotic pressure between the drops and the medium, otherwise the drops would not remain intact but would diffuse into the solution. Since under these conditions the liquid of the medium around the drops is perfectly symmetrical and homogeneous, it cannot exercise any influence on the liquid of the drops.

It is otherwise, however, with the colouring matter of the {81} drops. The particles of Indian ink may be seen passing from one drop to another, the coloured circles become elongated towards one another, touch, and finally unite. If, as in Fig. 23, the drops are of different size, the larger one will have a preponderating attractive action and eat up the smaller drops. In the figure, six small drops are placed around a large one, and the smaller drops have begun to be deformed and to move towards the larger drop. This central drop is also deformed, and has assumed a more or less hexagonal form, under the influence of the attraction of the six smaller ones. It may be noticed that the least prominent angle of the hexagon is opposite the small drop which is farthest away from it, whilst one of the smaller drops has already begun to be swallowed up by the large one. This cohesion phenomenon is very slow in its action, but after an hour or two the central drop will be found to have {82} completely absorbed the six smaller ones, and only one large drop will remain.

Incubation.--In the living organism we frequently find conditions similar to those realized in this experiment, viz. very slow movements of diffusion in liquids containing particles in suspension. In such cases the consequences must be the same, viz. granulation and segmentation. Consider for a moment the incubation of an egg. The heat of incubation determines a certain amount of evaporation through the shell, with a concentration of the liquid near the surface. As a consequence of this superficial concentration we get segmentation of the vitellus, with the production of a morula.

Artificial Parthenogenesis.--The experimental parthenogenesis of Loeb and Delage consists in plunging the egg into a liquid other than sea water, and returning it again to its original medium. This operation will necessarily determine slow movements of diffusion in the egg, which will give rise to segmentation. It may be objected that segmentation is also produced by a solution which is isotonic with sea water. Such a solution would not indeed produce an exchange of water with the egg, but it would set up an exchange of electrolytes, since there would be a difference of their osmotic pressure in the egg and in the new isotonic medium. The extremely slow movements of diffusion thus produced would be very favourable to the action of the cohesive force on the particles in suspension, and hence to the segmentation of the egg.

Few physical phenomena give us a deeper insight into the phenomena of life than those which we here contemplate. There is still another experiment which is even more convincing. On the surface of our horizontal salt solution we sow a number of drops of a more concentrated salt solution at equal distances around the circumference of a circle. Movements of diffusion are thus set up in the interior of the circle, and after a time, when this diffusion has become so slow as to be almost imperceptible, a furrow begins to appear in the coloured mass. Then a second and third appear, and others crossing the former break up the mass {83} into segments. Finally the segmentation becomes complete, and the preparation presents a muriform appearance, looking in fact something like a mulberry (Fig. 24). If the preparation is preserved for several hours longer, we may see the cells formed by segmentation unite around the circumference so as to form a hollow bag corresponding to a gastrula, as shown in Fig. 25.

These preparations are extremely sensitive to external influences, which renders the demonstration of cohesion phenomena difficult. I have nevertheless on several occasions been able to project the experiment on the screen during a lecture. The segmentation is influenced by very slight currents of diffusion, and I have many preparations showing the segmentation regularly distributed in various ways along radial diffusion lines. We may in this way produce many varieties of structure lamellar, vacuolate, or cellular, in fact {84} all the tissue structures which are met with in living organisms. All these structures are retractile, the retraction going on very slowly for a long time, as if the force of cohesion continued to act in the web of the structure even after its formation was complete. The phenomenon is a purely physical synthetic reproduction of the phenomenon of coagulation, the cohesion figure being in fact a retractile clot.

Crystallization.--When we evaporate a solution of a crystalloid it becomes more concentrated, slow movements of diffusion are set up, and at a given moment agglomeration occurs, the agglomerates taking the form of crystals. Thus crystallization may be regarded as a particular case of conglomeration by cohesion, differing only in the regularity of the arrangement of the molecules, which gives the geometrical form of the crystal. Hence we can easily understand how the presence of a crystalline fragment may facilitate the process of crystallization. Consider a liquid in which extremely slow movements of diffusion are taking place. If the liquid is perfectly homogeneous there will be no centre of attraction to which the molecules may become attached. {85}

If, however, a crystal or other heterogeneous structure is present, it forms a centre of cohesion which will attach any molecules that are brought by diffusion into its sphere of attraction. We have succeeded in photographing the arrangement of the molecules of a liquid around a crystal in the act of formation (Fig. 26). For this purpose we add to the solution traces of some colloidal substance, such as gelatine or gum, so as to delay the crystallization. It may thus be shown that the molecules of the surrounding liquid are already arranged in crystalline order for some distance from the crystal, forming a sort of field of crystallization. The arrangement of this regular field varies in different cases, and is more or less complicated according to circumstances. One of the most frequent forms is that shown in Fig. 27, which is the field around a crystal of sodium chloride. In the centre {86} of the crystal is a square with well-marked outline. At each corner of this square there is a straight line at right angles to the diagonal, which will form the sides of the crystal in process of formation. From the middle of each side arise yet other perpendiculars, which in their turn bear other cross lines, each new line being set at right angles to its predecessor. A later stage of crystallization is shown in Fig. 27, where the two squares one inside the other at an angle of 45?are clearly indicated.

Every crystallizable substance gives a different characteristic field of crystallization. In 1903, at the Congress of Angers, I terminated my address by these words: "The field of crystallization may serve to determine the character of a substance in solution."

Six years ago I received from Australia an exceedingly beautiful photograph of a thin pellicle found in a rain gauge. My correspondent supposed that this strange figure might have been produced under the influence of an electric or magnetic field. I was able to assure him by return of post that the figure was the result of the crystallization of copper sulphate in a colloidal medium. In return I received a letter verifying this fact, and saying that there were copper works in the neighbourhood, and the air was filled with the dust of copper sulphate.

Living beings are but solutions of colloids and crystalloids, and their tissues are built up by the aggregation of these solutes. We have already seen how the forces of crystallization are modified in colloid solutions. This force of crystallization must play an important role in the metamorphoses of the living organism, and influence their morphology. It may therefore be of interest to investigate some of the numberless forms of crystallization in colloidal solutions.

Figs. 29 and 30 represent the forms produced by chloride of sodium and chloride of ammonium respectively, in solutions of gelatine of different degrees of concentration. Their resemblance to vegetable growth is so remarkable that several observers on first seeing them have called them "Fern-crystals."

I should like here to recall to your notice the work of an English observer, Dr. E. Montgomery of St. Thomas's {88} Hospital, which was published as long ago as 1865. This work was recently brought to my notice by the kindness of Professor Baumler of Freiburg. He says: "Crystals are not strangers in the organic world. Many organic compounds are able to assume crystalline forms under certain conditions. Rainey has shown that many shells consist of globular crystals i.e. of mineral substances made to crystallize by the influence of viscid material." In this connection I may also mention the interesting work of Otto Lehmann of Karlsruhe on liquid crystals.

In conclusion, we may recall the words of Schwann himself, the originator of the cell theory: "The formation of the elementary shapes of an organism is but a crystallization of substances capable of imbibition. The organism is but an aggregate of such imbibing crystals."

* * * * *

{89}

CHAPTER VIII

KARYOKINESIS

In 1873, Hermann Fol, writing of the eggs of Geryonia, thus describes the phenomenon of karyokinesis: "On either side of the residue of the nucleus there appears a concentration of plasma, thus forming two perfectly regular star-like figures, whose rays are straight lines of granulations. There are other curved rays which pass from one star or centre of attraction to the other. The whole figure is extraordinarily distinct, recalling in a striking manner the arrangement of iron filings surrounding the poles of a magnet. Sachs' theory is that the division of the nucleus is caused by centres of attraction, and I agree with him, not on theoretical grounds, but because I have actually seen these centres of attraction."

Since the discovery of Hermann Fol, a great number of explanations have been given, all of them theoretical, to account for the figures and phenomena of karyokinesis. Many of these so-called explanations are mechanical, while others invoke the aid of magnetism or electricity to account for the resemblance of the figures of karyokinesis to the magnetic or electric phantom or spectre. Among the authors who have dealt with this question we may mention Hartog of Cork, Gallardo of Buenos Ayres, and Rhumbler of Gastingen.

In 1904 I presented to the Grenoble Congress, and in 1906 to the Lyons Congress, a series of photographs and preparations of experimental karyokinesis. I showed how, in a solution analogous to that found in the natural cell, the simple processes of liquid diffusion, without the intervention of magnetism or electricity, may reproduce with perfect accuracy and in their normal sequence the whole of the movements and {90} figures which characterize the phenomenon of karyokinesis. This experiment consists not merely in the production of a certain figure, such as is obtained in the magnetic spectre, but in the reproduction of the movement itself, and of all the successive forms which are seen in the natural phenomenon. These are evolved before the eyes of the spectator in their regular order and sequence.

I may here reproduce the text of my communication at Grenoble: "Until I introduced the conception of a field of diffusion, there was no proper means of studying the phenomena of diffusion, which obey the laws of a field of force as expounded by Faraday. Moreover, no one suspected the possibility of reproducing by liquid diffusion a spectre analogous to the electro-magnetic phantom. Guided by this theory of a diffusion field of force, I have been able to reproduce experimentally the figures of karyokinesis by simple diffusion. With regard to the achromatin spindle, Professor Hartog has shown that the two poles of the spindle are of the same sign, and not of opposite signs as was at first supposed. In the process of karyokinesis the two centrosomes, i.e. the two poles of the achromatin spindle, repel one another. They must therefore be poles of the same sign. An electric or magnetic spectre showing a spindle between two poles of the same sign is unknown; such a thing would appear to be an absolute impossibility. What is impossible in electricity and magnetism, however, is quite possible in the artificial diffusion field; we can here have a spindle between two poles which repel one another--that is, between poles of the same sign. Fig. 31 is a photograph of such a spindle produced by diffusion. On either side are two poles of concentration, which represent the centrosomes, each pole being surrounded by a star-like radiation. These poles being alike, repel one another. In the preparation one may see the distance between the two poles slowly increase, the poles gradually separating from one another just as do the centrosomes of an ovum during karyokinesis. This preparation, then, which is produced entirely by diffusion, presents a perfect resemblance to the achromatin spindle in karyokinesis. {91}

"The spindle of which we give a photograph in Fig. 31 was made by placing in salt water a drop of the same solution pigmented with blood or Indian ink, and placing on either side of this central drop a hypertonic drop of salt solution more lightly coloured. After diffusion had gone on for some minutes, we obtained the figure which we have photographed. I would draw your attention to the equatorial plane, which shows that the spindle is not formed by lines of force passing from one pole to the other, as would be the case between two poles of contrary sign, but by two forces acting in opposite directions. On either side the pigment of the central drop has been drawn towards the hypertonic centre nearest to it. In the median line, however, the pigment is attracted in opposite directions by equal forces, and therefore remains undisturbed, marking the position of the equatorial plane. This observation applies equally to the equatorial plane in natural karyokinesis, whose existence is thus readily explained.

"It is hardly necessary to insist on the fact that liquid preparations like these are of extreme delicacy and sensitiveness, and require for their production, and still more for their photography, the greatest care and skill, which can only be acquired by long practice. {92}

"We are able to produce by diffusion not only the achromatin spindle, but also the segmentation of the chromatin, and the division of the nucleus. If in the saline solution we place a coloured isotonic drop between two coloured hypertonic drops, all the figures and movements of karyokinesis appear successively in their due order. The central drop, representing the nucleus between the two lateral drops or centrosomes, first becomes granular. Next we see what appears to be a rolled-up ribbon analogous to the chromatin band, which soon breaks into fragments analogous to the chromosomes. These arrange themselves around, and are gradually attracted towards the centrosomes, where they accumulate to form two pigmented nuclear masses. A partition then makes its appearance in the median line, and this partition becomes continuous with the boundary of the spheres around the centrosomes. Finally we have two cells in juxtaposition, each with its nucleus, its protoplasm, and its enveloping membrane. I have been able to photograph these successive stages of the segmentation of the chromatin just as I have those of the achromatin spindle" (Fig. 32).

This memoir, written in 1904, clearly asserts the homopolarity of the centrosomes, and shows that the nuclear division is the result of a bipolar action, two poles of the same sign exerting their influence on opposite sides of the nucleus. It also emphasizes the important fact that diffusion, {93} and as far as we know diffusion alone, is able to produce a spindle between homologous poles.

A glance at the photograph is enough to show that the spindle is formed between poles of the same sign. The lines of diffusion radiate from one centre and converge towards the other centre in curves, giving the double convergence characteristic of a spindle. The central drop merely supplies the necessary material, and should have a concentration but slightly less than that of the plasma, so as not to set up its own lines of diffusion. The photograph shows clearly that the rays of the spindle traverse the equator without any break. It has been objected that these lines form not so much a spindle as two hemi-spindles, but it is clear that these two hemi-spindles are continuous and form a single sheaf of rays uniting the two poles of concentration. This is a phenomenon entirely unknown in the magnetic or electric fields, where two poles of the same sign, one on either side of a pole of the contrary sign, give two separate spindles. In a magnetic field it is impossible to make the lines emanating from one pole converge, except to a pole of opposite sign. Hence if we admit the homopolarity of the centrosomes, we must also admit that diffusion is the vera causa of karyokinesis, since, as I showed at the Grenoble Congress in 1904, diffusion and diffusion alone is capable of producing a spindle between two poles of the same sign.

Nuclear Division.--In order to reproduce artificially the phenomena attending the division of the nucleus, we may proceed as follows. We cover a perfectly horizontal glass plate with a semi-saturated solution of potassium nitrate to represent the cytoplasm of the cell. The nucleus in the centre is reproduced by a drop of the same solution coloured by a trace of Indian ink, the solid particles of which will represent the chromatin granules of the nucleus. The addition of the Indian ink will have slightly lowered the concentration of the central drop, and this is in accordance with nature, since the osmotic pressure of the nucleus is somewhat less than that of the plasma. We next place on either side of the drop which represents the {94} nucleus a coloured drop of solution more concentrated than the cytoplasm solution. The particles of Indian ink in the central drop arrange themselves in a long coloured ribbon, apparently rolled up in a coil, the edges of the ribbon having a beaded appearance. After a short time the ribbon loses its beaded appearance and becomes smooth, with a double outline, as is shown in A, Fig. 32. This coil or skein of ribbon subsequently divides, forming a nuclear spindle, while the chromatin substance collects together in the equatorial plane as in B, Fig. 32.

A more advanced stage of the nuclear division is shown at C, Fig. 32, where the chromatin bands of artificial chromosomes are grouped in two conical sheafs converging towards the two centrosomes. For some considerable time these conical bundles remain united by fine filaments, the last vestiges of the nuclear spindle. The final stage is that of two artificial cells in juxtaposition, whose nuclei are formed by the original centrosomes augmented by the chromatin bands or chromosomes (Fig. 32, D).

The resemblance of these successive phenomena to those of natural karyokinesis is of the closest. The experiment shows that diffusion is quite sufficient to produce organic karyokinesis, and that the only physical force required is that of osmotic pressure. If in the cytoplasm of a cell there are two points of molecular concentration greater than that of the general mass, the nucleus must necessarily divide with all the phenomena which accompany karyokinesis. In nature these two centres of positive concentration are introduced into the protoplasm of the cell by fecundation--that is, by the entrance of the centrosomes of the sperm cell. In certain abnormal cases the concentration may be produced in the cell itself by the formation of two centres of catabolism or molecular disintegration, since, as we have seen, molecular disintegration raises the osmotic pressure. This phenomenon, namely the production of karyokinesis from centres of catabolism, may account for the abnormal karyokinesis of cancer cells and the like. The subject is one which would well repay further investigation. {95}

[Illustration: FIG. 34.--A triaster produced by diffusion.]

It has been found in our experiments that in order to obtain the regular division of the artificial nucleus represented by the intermediary drop, the latter must have an osmotic pressure slightly below that of the plasma. This leads to the supposition that a similar condition must obtain in the natural cell. It may be noticed, moreover, that the grains of pigment follow the direction of the flow of water, being carried along by the stream. This would appear to show that the nucleus of a natural cell has also a molecular concentration less than that of the plasma--a result either of dehydration of the plasma, or of some diminution in the molecular concentration of the nucleus.

Other phenomena of karyokinesis may also be closely imitated by diffusion. For instance, in the diffusion preparation we notice at each extremity of the equator a V-shaped figure with its apex towards the centre, corresponding exactly to what in natural karyokinesis is called the equatorial crown.

We may also produce diffusion figures of abnormal karyokinesis. Fig. 34 represents such a form, a triaster produced by diffusion.

Artificial karyokinesis may also be produced by hypotonic poles of concentration--that is to say, when the central drop representing the ovum is positive and the lateral drops representing the centrosomes are negative with respect to the plasma. In this case, however, the resemblance to natural karyokinesis is less perfect. {96}

Without attaching to it an importance which is not warranted by experimental results, it is interesting to note that we have here two methods of fertilization, hypertonic and hypotonic, i.e. by centrosomes of greater concentration and by centrosomes of less concentration than that of the plasma of the ovum, and that we have in nature two corresponding results, viz. two different sexes. It is possible that we have in these two methods of producing nuclear division the secret of the difference of sex.

* * * * *

{97}

CHAPTER IX

ENERGETICS

Movement is everywhere; there is no such thing as immobility; the very idea of rest is itself an illusion. Immobility is only apparent and relative, and disappears under closer examination. All terrestrial objects are driven with prodigious velocity around the sun, and the dwellers on the earth's equator travel each day around the 40,000 kilometres of its circumference. All objects on the globe are in motion, the inanimate as well as the living. The waters rise in vapour from the sea, float over mountain and valley, and return down the rivers to the sea again. Still more marvellous is the current of water which flows eternally from dew and rain, through the sap of plants and the blood of animals to the mineral world again. The very mountains crumble and their substance is washed down into the plains; the winds move the air and raise the waves of the sea, whilst the strong ocean currents are produced by variations of temperature in different parts. This agitation, this incessant and universal motion, has been a favourite subject of poetic contemplation. Heraclitus writes: "There is a perpetual flow, all is one universal current; nothing remains as it was, change alone is eternal." Ovid writes in his Metamorphoses: "Believe me, nothing perishes in this vast universe, but all varies, and changes its figure. I think that nothing endures long under the same appearance. What was solid earth has become sea, and solid ground has issued from the bosom of the waters."

It was only towards the middle of last century that mankind in the long search after unity in nature began to realize that all the movements of the universe are the manifestations of a single agent, which we call energy. In reality all the phenomena of nature may be conceived as diverse forms of motion, and the word "energy" is the common expression applied to all the various modes of motion in the universe. It was by the study of heat, and more especially of thermodynamics, that we obtained our conceptions of the science of energetics.

It was in Munich in 1798 that the English engineer Count Rumford first observed that in the operation of boring a cannon the copper was heated to such a degree that the shavings became red-hot. This suggested his famous experiment, in which a heavy iron pestle was turned by horse power in a metal mortar filled with water. The water boiled, and when more water was added this also became heated to ebullition, and so on indefinitely. Rumford argued that the heat thus obtained in an indefinite quantity could not be a material substance; that motion was the only thing added to the water without limit, and that therefore heat must be motion.

While Rumford's experiment showed the transformation of motion into heat, the steam engine was soon afterwards to demonstrate the opposite transformation, viz. that of heat into motion.

The actual state of our knowledge with regard to the science of energy rests on two principles, that of Mayer and that of Carnot.

The first principle was defined by J. R. Mayer, a medical practitioner of Heilbronn, whose work, Bemerkungen ueber die Krte der unbelebten Natur, was published in 1842. "All physical phenomena," says Mayer, "whether vital or chemical, are forms of motion. All these forms of motion are susceptible of change into one another, and in all the transformations the {99} quantity of mechanical work represented by different modes of motion remains invariable."

The energy of a given body is the amount of transferable motion stored up in that body, and is measured by its capacity of producing mechanical work.

Ostwald thus defines energy: "Energy is work, all that can be obtained from work, and all that can be changed into work." Different forms of energy may be measured in different ways, but all forms of energy can be measured either in units of mechanical work or in units of heat, in kilogramme-metres or foot-pounds or in calories, according as the energy in question is transformed into mechanical work or into heat. The first principle of energetics, the conservation of energy, may be thus expressed: "Energy is eternal; none is ever created, and none is ever lost. The quantity of energy in the universe is invariable, and is conserved for ever in its integrity."

The unit by which we measure quantities of heat is the calory, the amount of heat required to raise the temperature of one kilogramme of water one degree Centigrade.

The practical unit of mechanical work is the kilogramme-metre, the work required to raise the weight of one kilogramme to the height of one metre. The theoretical unit of work is one erg, the work required to move a mass of one gramme through one centimetre against a force of one dyne.

Joule of Manchester was the first to verify Mayer's law quantitatively. By an experiment analogous to that of Rumford, he transformed work into heat, arranging his apparatus so that he might measure the amount of heat produced and the work expended. On dividing the quantity of work that had disappeared by the quantity of heat which had been disengaged, he found that 424 kilogramme-metres of work had been expended for each calory of heat produced.

Hirn of Colmar measured the ratio of work to heat in the steam engine. He found that for each calory of heat which had disappeared there were produced 425 kilogramme-metres of work. {100}

This number 425 has therefore been accepted as representing in calories and kilogramme-metres the transformation of work into heat, and of heat into work.

Further measurements on the transformations of other forms of energy, chemical energy and electrical energy, have shown that Joule's law of equivalents is general, and that the quantity of mechanical work represented by any form of energy remains undiminished after transformation, whatever the nature of that transformation.

Energy presents itself to us under two forms, potential and actual. Potential energy is slumbering energy, energy localized or locked up in the body. In order to transform potential energy into actual energy, there is required the intervention of an additional awakening, stimulating, or exciting energy from without. This stimulating energy may be almost infinitesimal in amount and bears no quantitative relation to the amount of energy transformed. It is the small amount of work required to turn the key which liberates an indeterminate quantity of potential energy.

Actual energy, on the other hand, is energy in movement, awake and alert, ready to be transformed into any other form of energy without the intervention of any such external stimulating force.

The passage of a given quantity of energy from the potential into the actual state is effected gradually, and during the time of transformation the sum of the actual and the potential energy remains constant.

A weight suspended by a cord possesses a quantity of potential energy equal to the product of its weight into the height through which it can fall.

This energy is locked up in a certain space, it cannot be transformed without the intervention of some external energy to cut the cord. During the falling of the weight, at the middle of its path, half of this slumbering energy has become kinetic, and is represented by the vis viva of the weight, while the other half is still potential and is equivalent to the work which the weight will accomplish during the second half of its fall. At any moment the sum of these two energies, the sleeping and the waking {101} energies, represents the total potential energy of the weight before it began to fall.

So with the powder in a gun. The potential energy of the powder cannot become actual without some stimulus, some exciting force from without to set it free. It is the external work of pressing the trigger that liberates the potential energy of the powder, transforming it into the actual energy of combustion, and the kinetic energy of the projectile.

Since energy is work, and work is a function of motion, there is in reality no such thing as energy in repose. Matter according to our modern conception is a complex of molecules, atoms, and electrons; we conceive the molecules of matter as always in movement, animated with cyclic or vibratory motion, these oscillatory or rotatory movements representing the potential energy of the body in question. Potential energy is thus the expression of molecular motion without translation of the molecules as a whole in space.

When this potential energy is transformed into actual energy by the intervention of some external force, we get a current of energy, a transference of the molecules in space. Thus, when an external force has released the weight, the molecular orbits in the falling body change in form, and the potential energy of the molecular motion becomes the kinetic energy of the falling body. Similarly in the conduction of heat, the energy of the hot body is transferred to a colder body by transmission of the vibratory motion from molecule to molecule. So again with chemical energy, the molecular motion of combustion may be transformed into the radiant energy of the ethereal waves.

Actual energy may be regarded as a current of molecular motion. To make the matter clearer, let a mass of matter be represented by a regiment of soldiers. Then each soldier will represent an electron, a company will be an atom, and a battalion will be a molecule. As long as the soldiers mark time, turn, or otherwise exercise without advancing, we have simply an accumulation of potential energy. The word of command, "March," is the exciting force which suddenly transforms this potential into kinetic energy. The marching {102} regiment is a representation of a body possessing kinetic energy. Potential energy is energy confined to a certain point in space, whereas actual energy is a current of energy, continually changing its place or form. Energy is like water-power--potential in the lake, actual in the waterfall or river.

Any mechanism capable of causing one form of energy to pass into another is a transformer of energy. A steam engine is a transformer of energy, changing caloric energy into mechanical work. An electrical machine is a transformer of energy, converting mechanical motion into a current of electricity, whilst an electro-motor changes the movement of electrons into mechanical movement. Every living being, and even man himself, is but a transformer of energy, changing the energy derived from the earth and air and sun into mechanical motion, nervous energy, and heat.

The first law of energetics, that of the conservation of energy, is analogous to Lavoisier's principle in chemistry, the conservation of matter. The sign of equality which unites the terms of a chemical equation expresses the fact that after every chemical reaction the same total mass of matter is present as before the transformation. This is also true of energy; after every transformation we find exactly the same total quantity of energy as before it. This, however, tells us nothing as to the conditions of the transformation, or the causes, i.e. the anterior phenomena, which determined such transformation.

The second principle of energetics, that of Carnot, enunciated in 1824, deals with the conditions under which a transformation of energy is possible. A mass of water at a certain height represents a quantity of potential energy equal to the product of its weight by its height; but this energy cannot produce mechanical work unless the water is allowed to fall. Consider two lakes at the same altitude and of the same capacity, one of which is entirely landlocked, while the other has an open channel leading to the sea. Each lake represents the same quantity of potential energy, but the energy of the landlocked lake is useless, it cannot be {103} transformed; whereas the other lake whose water can run into the sea realizes the conditions necessary for utilization, viz. the transformability of its energy. The same may be said of all forms of energy; a heat engine can only act as a transformer, change heat into work, if there is a difference of temperature between its source and its sink; an electric motor can only work if there is a fall of potential between the entrance and the exit of the electric current.

Energy presents itself to us as the product of two factors, weight and height in the waterfall, quantity and temperature in the heat engine, current intensity and potential in the electric motor.

In considering these two factors we may note that one factor is always a quantity (Q) and the other an intensity (I). This latter expresses some sort of difference of position or condition, the height of the weight, a difference of temperature in the heat engine, of pressure in the gas engine, or of electric potential in the dynamo or electric furnace. There can be no current of energy without this difference of potential, and therefore no transformation from one form of energy to another.

The second law of thermodynamics, Carnot's law, may therefore be enunciated thus: "Energy cannot be transformed without a fall of potential."

We may also derive this principle from a consideration of the formula of efficiency, the ratio of the work done by the transformer to the work done on the transformer.

Efficiency = energy transformed / total energy absorbed

The total energy is the product QI, i.e. the product of the total quantity by the total intensity at our disposal. The transformed energy is Q(I - I'), the product of the total quantity by the difference of intensity at the inlet and at the outlet of the machine. The formula for efficiency thus becomes

$$Q(I - I') / QI = (I - I') / I.$$

If I represents a temperature, then in order that the efficiency may be positive I' must be less than I, {104} there must be a fall of temperature in the machine. If I' were greater than I, i.e. if the temperature at the outlet were greater than that at the inlet, the efficiency would be a negative one, and the transformer would have to borrow heat from some external source.

Entropy.--In every transformation of energy a certain portion of the energy is transformed into heat: a lamp gives out useless heat as well as light, a machine gives out useless heat as well as mechanical work. This loss of useful energy as heat occurs in every transference or transformation of energy; it is only in the case of heat passing from a hotter to a colder body that there is no such transformation. When equality of temperature is established there has been no loss of energy, but the whole of the energy has become unutilizable, i.e. untransformable. In the formula of efficiency the fall of intensity I - I' is now zero, and therefore the efficiency of the machine

$$(I - I') / I$$

is also zero.

Since in all its transformations a certain fraction of the energy is changed into heat, there is a tendency in nature for all differences of temperature to become equalized. Hence the quantity of utilizable energy in the universe tends to diminish. Clausius called this unutilizable energy enmeshed in the substance of a body its entropy, and showed that in every transformation the amount of this unutilizable energy tended to increase. "The entropy of a system always tends towards a maximum value."

If this gradual incessant increase of entropy is universal in nature, and if there is no compensatory mechanism, the universe must be tending towards a definite end, when the whole of its energy shall have been transformed into unutilizable heat with a uniform temperature. There is, however, reason to suppose that some such compensatory mechanism does in fact exist. Behind us stretches an infinite past, and in the future we believe that the phenomena of nature will be unrolled in a cycle which has no end. But the arguments derived from a study of entropy apply only to the facts and phenomena actually under our notice, the supposed {105} impossibility, without borrowing energy from without, of re-establishing the differences of temperature by drawing heat from a colder in order to concentrate it in a hotter body, and may not be absolutely identical with those obtaining in other ages. Our ignorance of such a phenomenon and our powerlessness to produce it in no way argue that it is impossible. It may exist for aught we know in some other region of space, or in another time than ours. We may perhaps some day obtain artificially the conditions which would render possible such a phenomenon, since it may be possible to produce in the experimental laboratory conditions which are not spontaneously realized in nature under present conditions. The future may perchance reveal to us absolutely new phenomena which have not hitherto been realized. In his work on the evolution of matter and of energy Gustave le Bon gives expression to some interesting and original ideas on this subject.

The laws of Mayer and Carnot alone are not sufficient to explain the phenomena of life, without some consideration of the laws of stimulus. Mayer's principle asserts the conservation of energy, and Carnot's the conditions necessary for its transformation, but these alone cannot account for the transformation of potential into actual energy. A weight suspended by a cord does not fall merely because there is room for its descent. We need the intervention of some outside force to cut the cord. In every transformation of energy this external force is required to cut the cord, or pull the trigger, some external force of excitation or liberation, an energy which may be infinitesimal in amount, and which bears no proportion to the quantity of potential energy it sets free. This intervention of an excitatory, stimulating, or liberating energy is universal. Every phenomenon of nature is but a transformation or a transference of energy, determined by the intervention of a minimal quantity of energy from without. This liberation of large quantities of potential energy by an exceedingly small external stimulus has not hitherto received the consideration it demands. Certain phenomena, such as those of chemical catalysis or the action of soluble ferments, excite our astonishment because such extremely small quantities of {106} certain substances will determine the chemical transformations of large quantities of matter, there being no proportion between the amount of the catalytic substance and of the matter transformed. These phenomena are, however, only particular cases of the general law of energetics that transformation requires a stimulus. The catalyzer, or ferment, does not contribute matter to the reaction, but only the minimal energy necessary to liberate the chemical potential energy stored in the fermenting substance.

We must therefore add a third to the two laws of energetics, Mayer's law of conservation, and Carnot's law of fall of potential. This third law is the law of stimulus, the necessity of the intervention of an external excitatory force capable of setting in motion the current of energy required for a transformation. This stimulus is the primary phenomenon, the determinant cause of such transformation.

Three conditions, then, are required for a transformation or displacement of energy:--

1. The cause, the intervention of a stimulus which starts the transformation or displacement.

2. The possibility, the necessary fall of potential.

3. The condition, the conservation of the energy concerned, since being indestructible its total quantity cannot alter.

Every living being is a transformer of energy. The lower animals and man himself receive from food and air the potential energy which becomes actual under the process of oxydation. This chemical combustion is the source of all vital energy; the ancients aptly compared life to a flame, and Lavoisier has shown that life, like the flame, is maintained by a process of oxydation. The energy derived from food and air is restored by the organism to the external world in the form of heat and mechanical motion. The celebrated experiments of Atwater show that there is an absolute equality between the energy obtained from the oxydation of the various aliments and the sum of the calorific and mechanical energy liberated by a living being.

Man obtains his supply of energy either directly from the {107} vegetable world, or indirectly from vegetables which have passed through the flesh of animals. Vegetables in their turn obtain their substance from the mineral world and their energy from the sun. The salts, the water, and the carbonic acid absorbed by plants possess no store of potential energy. Whence then can they obtain the potential energy which they transmit to animals and man, if not from the sun? The energy of the solar radiations is absorbed by the chlorophyll of the leaves, and stored up in the organic carbohydrates formed by the synthesis of water and carbon. Chlorophyll has the peculiar property of reducing carbonic acid, and uniting the carbon with water in different proportions to form sugar and starch, whilst fats and vegetable albumens are also formed by an analogous reaction. All these complex bodies are stores of energy; the vital processes of oxydation do but liberate in the human body the energy which the chlorophyll of plants has absorbed from the solar rays.

We must look, then, to the sun as the direct source of all the energy which animates the surface of the earth. The sun looses the winds, and raises the waters of the sea to the mountain-tops, to form the rivers and torrents which return again to the sea; the sun warms our hearths, drives our ships, and works our steam engines. There is no sign of life or movement on our planet which does not come directly or indirectly from the solar rays.

It may be asked by what path does the chemical energy of the living organism pass into the mechanical energy of motion. It would appear that the intermediary step cannot be heat, as in the steam engine, since the necessary temperature would be quite incompatible with life.

The formula for the efficiency of a thermic transformer is

$$(T - T') / T,$$

the ratio of the difference of the absolute temperatures at the source and at the sink, to the absolute temperature at the source. Calorimetric measurements have shown that the efficiency of the human machine is about one-fifth, i.e. it can transform 20 per cent. of the energy absorbed. The ordinary temperature of muscle is 38?C., or 311?absolute. We have {108} therefore $(T - 311) / T = .20$, or $T = 388.75$?absolute, i.e. 115.75?C. Thus, in order to obtain an efficiency of 20 per cent. with an ordinary thermic transformer, having a temperature of 38?at the sink, we should need a temperature of over 115?C. at the source. Such a temperature would be quite incompatible with the integrity of living tissues, and we may therefore conclude that the human organism is not a heat engine.

We are indeed completely ignorant of the mode of transformation of chemical into kinetic energy in the living organism; we know only that muscular contraction is accompanied by a change of form; at the moment of transformation the combustion of the muscle is increased, and during contraction the stretched muscular fibre tends to acquire a spherical shape. It is this shortening of the muscular fibre which produces the mechanical movement. The step which we do not as yet fully understand is the physical phenomenon which intervenes between the disengagement of chemical energy and the occurrence of muscular contraction. Professor d'Arsonval supposes that this missing step is a variation in the surface tension of the liquid in the muscular fibre. The surface tension of a liquid is due to the unbalanced forces of cohesion acting on the surface layer of molecules. Under the attraction of cohesion the molecules within the liquid are in a state of equilibrium, being equally attracted in all directions, but those at the surface of the liquid are drawn towards the centre. The resultant of these attractive forces is a pressure normal to the surface, which is mechanically equivalent to an elastic tension tending to diminish the surface. In consequence of this surface tension the liquid has a tendency to assume the form in which its surface area is a minimum, i.e. the spherical form. If such a sphere is stretched into a cylinder or fibre by mechanical tension, it will shorten itself when released; and if by any means we increase the surface tension of such a liquid fibre it will tend to assume a spherical form and contract just as a muscular fibre does. The surface tension of a liquid varies with its chemical composition; the slightest chemical modification of a liquid alters the force of {109} this tension. We may therefore explain the mechanism of muscular contraction by supposing that a nervous impulse alters in some way the rate of combustion in a muscular fibre, that this alteration produces a momentary change in the chemical composition of the muscular cell, and that this change of chemical composition increases the surface tension of the cell sufficiently to provoke its contraction into a more spherical form.

Ostwald has introduced a very useful conception for the study of this question of surface energy. A liquid surface contains a quantity of energy equal to its surface tension multiplied by its area, hence any variation either of area or of tension corresponds to a variation of its energy. This novel conception constitutes a valuable addition to the experimental study of the physiology of muscular action, since it gives us some idea of the mechanism by which chemical energy may be transformed into muscular contraction.

Whatever the mechanism of transformation in the animal machine, we have to consider the same quantities as in other motor machines. These are: (1) the efficiency; (2) the potential energy; (3) the power; (4) the energy given up to the medium under the form of heat; (5) the temperature.

Muscles, then, are merely transformers which change chemical energy into mechanical work, the diminution of stored-up energy in a muscle being expressed by the sensation of fatigue. A muscle may be studied in four different phases: (1) in repose; (2) in a state of tension; (3) when doing positive work; (4) when work is being done on it.

When a muscle is in a state of tension, as when a weight is sustained by the outstretched arm, the muscle is producing no external work. The entire work done is converted into heat; just as it is in a dynamo or steam engine which is prevented from turning by a brake. Muscular contraction produces fatigue even when it does no external work. It is impossible for the muscle to support even the weight of the outstretched arm itself for any considerable time.

A muscle is doing positive work when it is raising a weight or moving a body from one point to another. {110}

The fourth state of muscular contraction is when the muscle is doing negative work, i.e. when work is being done on it, as for instance when we go downstairs, or when a descending weight forces down the opposing arm which attempts to support it. In this case the muscles receive a portion of the energy lost by the descending weight, and this energy shows itself in the muscle in the form of heat. This increase of heat in a muscle doing negative work has been clearly demonstrated by the calorimetric experiments of Hirn and the thermometric experiments of B 闰 lard. Hirn's observations on muscular calorimetry show a production of heat corresponding to 150 calories per hour when in repose, 248 calories per hour during positive work, and 287 during negative work. B 闰 lard's thermometric measurements also show that the temperature of a muscle rises each time that it contracts, and that the rise of temperature is greatest when the muscle is doing negative work, least during positive work, and intermediate when in a state of tension.

It is of the greatest importance in medical practice to distinguish between these different forms of muscular activity. There is a vast physiological difference between muscular contraction with the production of positive work, and muscular contraction without the production of work, or with negative work. To climb a flight of stairs is to contract the muscles with the production of work equal to the weight of the body multiplied by the height of the stairs. To descend the stairs is to contract the same muscles, but with the production of negative work, and consequently a maximum of heat. To walk on level ground is to contract the muscles with the production of little or no external work; as in a machine turning without friction in a vacuum.

We have seen that a fall of potential and a current of energy are the necessary conditions for the production of any natural phenomenon. Hence we may assume that the phenomenon of sensation is also accompanied by a fall of potential and a current of energy. When we touch a hot body, there is a flow of energy from the hot body to the hand. When we touch a cold body, there is a current of energy in the opposite direction, {111} from the hand to the body. It was formerly held, and is still held by some physiologists, that the chief characteristic of life is the disproportion between an excitation and the response which it invokes from the organism. Such a doctrine can only be held by one who believes, at least implicitly, that the phenomena of life are supernatural, or at all events different in their nature from all other phenomena; for the disproportion between an excitation and the response it evokes is by no means confined to living things. This disproportion is universal in nature, and quite in conformity with the physical laws which govern the transformation of energy. The energy of living things is potential energy--a fact which has been too little recognized. In the case of reflex actions it is self-evident, because the response is immediate, and always the same for the same stimulus. As in all other transformations, the stimulus consists in the intervention of a minimal quantity of external energy.

Long before the discovery of the laws of energy, Lamarck had recognized and formulated this fact. He writes: "What would vegetable life be without excitations from without, what would be the life even of the lower animals without this cause?" In another passage, seeking for a power capable of exciting the action of the organism, he says: "The lower animal forms, without nervous system, live only by the aid of excitations which they receive from without. In the lowest forms of life this exciting force is borrowed directly from the environment, while in the higher forms the external exciting force is transferred to the interior of the living being and placed at the disposal of the individual."

This remark, that the movements of living things are not communicated but excited, that the external excitation only sets free latent or potential energy in the organism, shows that Lamarck had penetrated more deeply than many of the modern physiologists into the secrets of biological energy. We seek in vain in the text-books of physiology for any conception of potential energy in living beings, or the notion of an exciting force as the cause of sensation. All action of a living organism is reflex action. Every action has a cause, and {112} the cause of an organic action is an exciting energy from without, either immediate, or stored up in the nervous system from an external impression made at some previous epoch. Actions which are not evidently reflex are merely delayed reflexes; we have acquired the power of inhibiting, delaying, or modifying the response to an external stimulus, so that the same excitation may determine responses of very different kinds according to the mood produced by previous impressions. When carefully investigated, no action of ours is automatic; every movement is determined by impressions derived from without. An action without a motive, that is without an external determining cause, would be an action without reason.

In conclusion, we may formulate this general principle: The energy of a living being is potential energy; sensations represent the intervention of an external exciting energy which provokes the response, i.e. the transformation of the potential energy already stored in the organism into the actual energy of motion and vital activity.

* * * * *

{113}

CHAPTER X

SYNTHETIC BIOLOGY

The course of development of every branch of natural science has been the same. It begins by the observation and classification of the objects and phenomena of nature. The next step is to decompose the more complex phenomena in order to determine the physical mechanism underlying them--the science has become analytical. Finally, when the mechanism of a phenomenon is understood, it becomes possible to reproduce it, to repeat it by directing the physical forces which are its cause--the science has now become synthetical.

Modern biology admits that the phenomena of life are physico-chemical in their nature. Although we have not as yet been able to define the exact nature of the physical and chemical processes which underlie all vital phenomena, yet every further discovery confirms our belief that the physical laws of life are identical with those of the mineral world, and modern research tends more and more to prove that life is produced by the same forces and is subject to the same laws that regulate inanimate matter.

The evolution of biology has been the same as that of the other sciences; it has been successively descriptive, analytical, and synthetic. Just as synthetic chemistry began with the artificial formation of the simplest organic products, so biological synthesis must content itself at first with the fabrication of forms resembling those of the lowest organisms. Like other sciences, synthetic biology must proceed from the simpler to the more complex, beginning with the reproduction of the more elementary vital phenomena. Later on we may hope to {114} unite and associate these, and to observe their development under various external influences.

The synthesis of life, should it ever occur, will not be the sensational discovery which we usually associate with the idea. If we accept the theory of evolution, then the first dawn of the synthesis of life must consist in the production of forms intermediate between the inorganic and the organic world--forms which possess only some of the rudimentary attributes of life, to which other attributes will be slowly added in the course of development by the evolutionary action of the environment.

Long ago, the penetrating genius of Lamarck seized on the idea that a knowledge of life could only be obtained by the comparison of organic with inorganic phenomena. He writes: "If we would acquire a real knowledge of what constitutes life, of what it consists, what are the causes and the laws which give rise to this wonderful phenomenon of nature, and how life can be the source of the multitude of forms presented to us by living organisms, we must before all consider with great attention the differences which exist between inorganic and living bodies; and for this purpose we must compare side by side the essential characters of these two classes of bodies."

Synthetic biology includes morphogeny, physiogeny, and synthetic organic chemistry, which is also a branch of synthetic biology, since it deals with the composition of the constituents of living organisms. Synthetic organic chemistry is already a well-organized science, important by reason of the triumphs which it has already gained. The other two branches of biological synthesis, morphogeny, the synthesis of living forms and structures, and physiogeny, the synthesis of functions, can hardly as yet be said to exist as sciences. They are, however, no less legitimate and no less important than the sister science of synthetic chemistry.

Although morphogeny and physiogeny do not exist as well-organized and recognized sciences, there are already a number of works on the subject by enthusiastic pioneers--independent seekers, who have not feared to abandon the paths of official science to wander in new and hitherto unexplored domains. {115}

The first experiment in physiogeny was the discovery of osmosis by the Abb?Nollet in 1748. He filled a pig's bladder with alcohol, and plunged it into water. He noticed that the bladder gradually increased in volume and became distended, the water penetrating into the interior of the bladder more quickly than the alcohol could escape. This was the first recorded experiment in the physics of nutrition and growth.

In 1866, Moritz Traube of Breslau discovered the osmotic properties of certain chemical precipitates. As I pointed out in the Revue Scientifique of March 1906, Traube made the first artificial cell, and studied the osmotic properties of membranes and their mode of production. This remarkable research should have been the starting-point of synthetic biology. The only result, however, was to give rise to numberless objections, and it soon fell into complete oblivion. "There are," says Traube, "a number of persons quite blind to all progress, who in the presence of a new discovery think only of the objections which may be brought against it." The works of Traube have been collected and published by his son (Gesammelte Abhandlungen von Moritz Traube, 1899).

In 1867 there appeared in England a paper by Dr. E. Montgomery, of St. Thomas's Hospital, On the Formation of so-called Cells in Animal Bodies. This paper, published by Churchill & Sons, is a most interesting contribution and one of great originality. The author says: "There can be no compromise between the tenets of the cell theory and the conclusions arrived at in this paper; the distinction is thorough. Either the units of which an organism is composed owe their origin to some kind or other of procreation, a mysterious act of that mysterious entity life, by which, in addition to their material properties, they become endowed with those peculiar metaphysical powers constituting vitality. Or, on the other hand, the organic units, like the crystalline units of inorganic bodies, form the organism by dint of similar inherent qualities, form in fact a living being possessed of all its inherent properties, as soon as certain chemical compounds are placed under certain physical conditions. If the former opinion be {116} true, then we must clearly understand that there exists naturally a break in the sequence of evolution, a chasm between the organic and the inorganic world never to be bridged over. If, on the contrary, the latter view be correct, then it strongly argues for a continuity of development, a gradual chemical elaboration, which culminates in those high compounds which, under surrounding influences, manifest those complex changes called vital.

"Surely it is not a matter of indifference or of mere words, if the extreme aim of physiology avowedly be the detection of the different functions dependent on the vital exertions of a variety of ultimate organisms, and the discovery of the specific stimulants which naturally incite these functions into play. Or, on the other hand, if it be understood to consist rather in the careful investigation of the succession of chemical differentiations and their accompanying physical changes, which give rise to the formation of a variety of tissues that are found to possess certain specific properties, to display certain definite actions due to a further flow of chemical and physical modifications."

In 1871 there appeared a memoir by the Dutch savant Harting entitled Recherche de Morphologie synthetique sur la production artificielle de quelques formations calcaires organiques. This memoir, says Professor R. Dubois, had cost Harting more than thirty years of work. "Synthetic morphology is yet only in its infancy, let us hope that in a time equal to that which has already expired since the first artificial production of urea, it will have made a progress equal to that of its older sister, synthetic chemistry."

In the Comptes Rendues of 1882 is the following note by D. Monnier and Karl Vogt:--

"1. Figured forms presenting all the characteristics of organic growth, cells, porous canals, tubes with partition walls, and heterogeneous granules, may be produced artificially in appropriate liquids by the mutual action of two salts which form one or more insoluble salts by double decomposition. One of the component salts should be in solution, while the other salt must be introduced in the solid form. {117}

"2. Such forms of organic elements, cells, tubes, etc., may be produced either in an organic liquid or a semi-organic liquid such as sucrate of lime, or in an absolutely inorganic liquid such as silicate of soda. Thus there can no longer be any question of distinctive forms as characterizing organic bodies in contradistinction to inorganic bodies.

"3. The figured elements of these pseudo-organic forms depend on the nature, the viscosity, and the concentration of the liquids in which they are produced. Certain viscous liquids such as solutions of gum arabic or chloride of zinc do not produce these forms.

"4. The form of these artificial pseudo-organic products is constant, as constant as that of the crystalline forms of mineral salts. This form is so characteristic that it may often serve for the recognition of a minimal proportion of a substance in a mixture. The observation of these forms is a means of analysis as sensitive as that of the spectrum. We may, for example, differentiate in this way the alkaline bicarbonates from the sesqui-carbonates or the carbonates.

"5. The form of these artificial pseudo-organic elements depends principally on the nature of the acid radical of the solid salt. Thus the sulphates and the phosphates generally produce tubes, while the carbonates form cells.

"6. As a rule these pseudo-organic forms are engendered only by substances which are found in the living organism. Thus sucrate of calcium will engender organic forms, whereas sucrate of strontium or barium does not do so. There are, however, some exceptions to this rule, such as the sulphates of copper, cadmium, zinc, and nickel.

"7. These artificial pseudo-organic elements are surrounded by veritable membranes, dializing membranes which allow only liquids to pass through them. These artificial cells have heterogeneous cell-contents, and produce in their interior granulations which are disposed in a regular order. Thus they are both in constitution and in form absolutely similar to the cellular elements which constitute living organisms.

"8. It is probable that the inorganic elements which are present in the natural protoplasm may play an important part {118} in determining the form which is assumed by the figured elements of the organism."

In 1902, Professor Quinke of Heidelberg, who has consecrated his life with such distinction to the physics of liquids, writes thus of the organogenic power of liquids in a paper published in the Annalen der Physik under the title "Unsichtbare Flesigkeitschichten": "In 1837, Gustav Rose obtained organic forms by precipitation from inorganic solutions. By precipitating chloride of calcium with the carbonates of ammonium and other alkaline carbonates, he obtained small spheres which grew and were transformed into calcic rhombohedra. He also obtained a flocculent precipitate which later became granular and showed under the microscope forms like the starfish, and discs with undulated borders. At Freiberg, in certain stalactites, Rose also discovered forms consisting of six pyramidal cells around a spherical nucleus.

"In 1839, Link obtained spherical granulations by the precipitation of calcic or plumbic solutions by potash, soda, or carbonic acid. These spherical granulations united after a time to form crystals. Sulphate of iron, ammoniated sulphate of zinc, sulphate of copper precipitated by sulphuretted hydrogen, and saline solutions precipitated by ferrocyanide of potash, all give granular precipitates or discs, of which the granular origin is quite perceptible.

"Runge in 1855 was the first to describe the formation of periodic chemical precipitates. He used blotting paper as the medium in which various chemical substances met by diffusion. In this way he studied the mutual reactions of solutions of ferrocyanide of potash, chloride of iron, and the sulphates of copper, iron, manganese, and zinc. The coloured precipitates appeared at different positions in the paper, and disappeared periodically at greater or longer intervals. The designs formed by these coloured precipitates change with the concentration of the saline solutions, or on the addition of oxalic acid, salts of potash or ammonia, and other substances. These designs are shown in a number of beautiful illustrations which accompany the work. In this {119} case the capillarity of the paper necessarily exerts a certain influence on the formation of the figures, but in addition to this, Runge admits the intervention of another force hitherto unknown, which he calls 'Bildungstrieb,' the formative impulse, which he considers to be the elementary vital force in the formation of plants and animals.

"In 1867, R. Bertger obtained arborescent forms and ramifications of metallic vegetation by sowing fragments the size of a pea of crystals of the iron chlorides, chloride of cobalt, sulphate of manganese, nitrate and chloride of copper, etc., in an aqueous solution of silicate of sodium of specific gravity 1.18. These forms are due, as I shall show later on, to the surface tension of the oily precipitate; Bertger gives no explanation of the phenomenon.

"To this force, viz. that of surface tension, is also due the cellular forms obtained by Traube in 1866. These were obtained from gelatine and tannin, from acetate of copper or lead, and from nitrate of mercury in an aqueous solution of ferrocyanide of potassium. These cells and precipitated membranes have also been studied by Reinke, F. Cohn, H. de Vries, and myself, who all observed the regression of these membranes, which although colloidal at the beginning of the reaction speedily become friable. This entirely refutes the opinion of Traube as to the constitution of the precipitated membranes. He supposed them to consist of masses of solid substance, with smaller orifices which do not permit the passage of the membranogenous substance, whilst the larger orifices through which it can pass are soon closed by the precipitate, the membrane itself thus growing by a process of intussusception.

"Later on Traube himself considered the precipitated membrane to be a thin, solid gelatinous layer in which the water was mechanically entangled.

"Tamman has also made a number of experiments with solutions of the chlorides and sulphates of the heavy metals, and solutions of phosphates, silicates, ferrocyanides, and other salts. He found that most of these membranes were permeable to the membranogenous solution. According to Tamman, all {120} precipitated membranes are hydrated substances, and some of them, like the ferrocyanide of copper and the tannate of gelatine are, when first formed, entirely comparable to liquid membranes in all their properties.

"Graham had already obtained colourless jellies by the interaction of concentrated solutions of ferrocyanide of potassium and sulphate of copper. Buschli also has recently described the microscopic appearance of precipitated membranes produced by ferrocyanide of potassium and acetate or chloride of iron.

"Like Linke and Gustav Rose, Famintzin has obtained spheroidal precipitates by the reciprocal action of concentrated solutions of chloride of calcium and carbonate of potassium. These grow rapidly and suddenly, with concentric layers showing a spherical or flattened nucleus. He also obtained forms resembling sphero-crystals and starch grains.

"Harting, Vogelsang, Hansen, Buschli, and others have studied the structures which are formed by the reciprocal action of chloride of calcium and the alkaline carbonates. Vogelsang has found small calcareous bodies in the amorphous and globular precipitate formed by chloride of calcium and carbonate of ammonium. He describes spheres attached to one another, vesicles, and muriform structures. The number of these spheroids is increased by the addition of gelatine. Hansen has also studied Harting's method for the formation of sphero-crystals by the action of the alkaline carbonates and phosphates on the salts of calcium in presence of albumen and gelatine. He considers that the latter retard the crystallization and assist the formation of the sphero-crystals.

The subjects of the numerous memoirs that I have myself published during the last ten years upon the question are treated anew in the pages of this volume, and a summaryof my researches on osmotic growth has already appeared in the Documents du Progres, Sept. 1909.

We have thus shown that synthetic morphogenesis has already attracted the attention of a certain number of ardent investigators. Morphogeny has now its methods and its results, and physiogeny is also developing side by side with it, since function is but the result of form. The field of research is opened, and workers alone are needed in order to reap an abundant harvest.

* * * * *

CHAPTER XI

OSMOTIC GROWTH--A STUDY IN MORPHOGENESIS

The phenomenon of osmotic growth has doubtless presented itself to the eyes of every chemist; but to discover a phenomenon it is not enough merely to have it under our eyes. Before Newton many a mathematician had seen a spectrum, if only in the rainbow; many an observer before Franklin had watched the lightning. To discover a phenomenon is to understand it, to give it its due interpretation, and to comprehend the importance of the role which it plays in the scheme of nature.

Osmotic Membranes.--Certain substances in concentrated solution have the property of forming osmotic membranes when they come in contact with other chemical solutions. When a soluble substance in concentrated solution is immersed in a liquid which forms with it a colloidal precipitate, its surface becomes encased in a thin layer of precipitate which gradually forms an osmotic membrane round it.

An osmotic membrane is not a semi-permeable membrane, as sometimes described, i.e. a membrane permeable to water but impermeable to the solute. It is a membrane which opposes different resistances to the passage of water and of the various substances in solution, being very permeable to water, but much less so to the different solutes.

A soluble substance thus surrounded by an osmotic membrane represents what Traube has called an artificial cell. In such a cell the dissolved substances have a very high osmotic pressure, an expansive force like that of steam in a boiler; the molecules of the solute exerting pressure on the walls of the extensible cell, and distending it like the {124} gas in a balloon. This pressure increases the volume of the cell, and in consequence water rushes in through the permeable membrane and still further distends the cell. Most beautiful osmotic cells may be produced by dropping a fragment of fused calcium chloride into a saturated solution of potassium carbonate or tribasic potassium phosphate, the calcium chloride becoming surrounded by an osmotic membrane of calcium carbonate or calcium phosphate. This mineral membrane is beautifully transparent and perfectly extensible. It is astonishing to contemplate the contrast between the hard crystalline forms of ordinary chalk and these soft transparent elastic membranes which have the same chemical constitution. These osmotic cells of carbonate of lime or phosphate of lime consist of a transparent membrane enclosing liquid contents and a solid nucleus of chloride of calcium. Their form is that of an ovoid or flattened sphere, and they may attain a diameter of seven centimetres or more.

More frequently the osmotic growth consists of a number of cells instead of one large cell. The first cell gives birth to a second cell or vesicle, and this to a third, and so on, so that we finally obtain an association of microscopic cellular cavities, separated by osmotic walls--a structure completely analogous to that which we meet with in a living organism.

We may easily picture to ourselves the mechanism by which an osmotic cell gives birth to such a colony of microscopic vesicles. The membranogenous substance, the chloride of calcium, diffuses uniformly on all sides from the solid nucleus, and forms an osmotic membrane where it comes into contact with the solution. This spherical membrane is extended by osmotic pressure, and grows gradually larger. Since the area of the surface of a sphere increases as the square of its radius, when the cell has grown to twice its original diameter, each square centimetre of the membrane will receive by diffusion but a quarter as much of the membranogenous substance. Hence, after a time, the membrane will not be sufficiently nourished by the membranogenous substance, it will break down, and an aperture will occur through which the interior liquid oozes out, forming in its turn a new {125} membranous covering for itself. This is the explanation of the fact that all living organisms are formed by colonies of microscopical elements, although we must not forget that Nature often produces similar results in different ways.

Osmotic growths may be obtained from a great number of chemical substances. The most easily grown are the soluble salts of calcium in solutions of alkaline phosphates and carbonates, to which we have already alluded. We may also reverse the phenomenon by growing phosphates and carbonates in solutions of calcium salts, but in this case the osmotic growths are not so beautiful.

The various silicates play an important part in the constitution of shells and of the skeletons of marine animals. Most of the metallic salts, and more especially the soluble salts of calcium, give rise to the phenomenon of osmotic growth when sown in solutions of the alkaline silicates. In this way, by using different silicates and varying the proportions and the concentrations, we may obtain an immense variety of osmotic growths.

A good solution to commence with is the following:--

Silicate of potash, sp. gr. 1.3 (33?Beaum? 60 gr. Saturated solution of sodium carbonate 60 gr. Saturated solution of dibasic sodium phosphate 30 gr. Distilled water make up to 1 litre.

{126}

A fragment of fused calcium chloride dropped into this solution will produce a rapid growth of slender osmotic forms which may attain a height of 20 or 30 centimetres.

Small pellets may also be made of one part of sugar and two of copper sulphate and sown in the following solution, which must be kept warm until the growth is complete:--

Ten per cent. solution of gelatine 10 to 20 c.c. Saturated solution of potassium ferrocyanide 5 to 10 c.c. Saturated solution of sodium chloride 5 to 10 c.c. Warm water (32?to 40?C.) 100 c.c.

In this solution we can obtain osmotic growths which may attain to a height of 40 centimetres or more, vegetable forms, roots, arborescent twigs, leaves, and terminal organs. These growths are stable as soon as the gelatine has cooled and set, and may be carried about without fear of injury (Fig. 35).

Precipitated osmotic membranes are very widely distributed in nature. Professor Ulenhuth has seen iron growths in alkaline sodium hypochlorite (Javelle water), and Lecha-Marzo has demonstrated the osmotic growth of the various {127} stains used for microscopy, in the liquids used for fixing preparations.

We now know that the physical force which builds up these growths is that of osmotic pressure, since the slightest consideration will show the inadequacy of the usual explanation that the growth is due to mere differences of density, or to amorphous precipitation around bubbles of gas. These may indeed affect the phenomenon, but can in no way be regarded as its cause.

One of our experiments throws considerable light on this question. In a glass vessel we placed a concentrated solution of carbonate of potassium, to which had been added 4 per cent. of a saturated solution of tribasic potassium phosphate. Into this solution we dropped a fragment of fused calcium chloride, and obtained a vermiform growth some 6 millimetres in diameter. This growth was curved, at first growing upwards, then for a short distance horizontally, and finally downwards. The upward pressure of the solution, which was heavier than the growth, ultimately broke it at the top of the curve, as shown at b, Fig. 37. The liquid contents of the growth began to ooze out through the wound, but this after a time became cicatrized, and the stem continued to grow obstinately downwards once more, in opposition to the hydrostatic pressure. In consequence of this pressure the growth is sinuous, tacking as it were from side to side like a boat against the wind. We give three successive photographs of this growth, which attained a length of over 10 inches. We have frequently obtained these vermiform growths forming a series of such loops, growing upwards and falling again many times in succession.

Osmotic Growths in Air.--Certain of these artificial cells may be made to grow out of the solution into the air. For this purpose we place a fragment of CaCl2 in a shallow flat-bottomed glass dish, just covering the fragment with liquid. The best solution is as follows:--

Potassium carbonate, saturated solution 76 parts. Sodium sulphate, saturated solution 20 " Tribasic potassium phosphate, saturated solution 4 "

The calcium chloride surrounds itself with an osmotic membrane; water penetrates into the interior of the cell thus formed, and a beautiful transparent spherical cell is the result, the summit of which soon emerges from the shallow liquid. The cell continues to increase by absorption of the liquid at its base, and may grow up out of the liquid into the air for as much as one or two centimetres.

This is a most impressive spectacle, an osmotic production, half aquatic and half aerial, absorbing water and salts by its base, and losing water and volatile products by evaporation from its summit, while at the same time it absorbs and dissolves the gases of the atmosphere.

The aerial portion of an osmotic growth will sometimes become specialized in form. The summit of the growth develops a sort of crown or cup surrounded by a circular wall. This cup contains liquid, and continues to grow up into the air like the stem of a plant, carrying with it the liquid which has been absorbed by the base of the growth.

The preceding experiments give us an explanation of the curious phenomena exhibited by so-called creeping salts. A saline solution left at the bottom of a vessel will sometimes be found after some months to have crept up to the top of the vessel. Cellular partitions formed in this way will be found extending from the bottom to the top of the vessel, and not only so, but the whole of the remaining liquid will be imprisoned in the upper cells.

Assimilation and Excretion.--Like a living being, an osmotic growth absorbs nutriment from the medium in which it grows, and this nutriment it assimilates and organizes. If we compare the weight of an osmotic growth with that of the mineral fragment which produced it, we shall find that the mineral seed has increased many hundred times in weight. Similarly, if we weigh the liquid before and after the experiment, we shall find that it has lost an equivalent weight. The absorbed substance of an osmotic production must also undergo chemical transformation before it can be assimilated--that is, before it can form part of the growth. Calcium chloride, for example, growing in a solution of potassium {130} carbonate, is transformed into calcium carbonate. $CaCl_2 + K_2CO_3 = CaCO_3 + 2KCl$. Thus an osmotic growth can make a choice between the substances offered to it, rejecting the potassium of the nutrient liquid, and absorbing water and the radical CO_3, while at the same time it eliminates and excretes {131} chlorine, which may be found in the nutrient liquid after the reaction.

Of all the ordinary physical forces, osmotic pressure and osmosis alone appear to possess this remarkable power of organization and morphogenesis.

It is a matter of surprise that this peculiar faculty has hitherto remained almost unsuspected.

Osmotic Growths.--If we sow fragments of calcium chloride in solutions of the alkaline carbonates, phosphates, or silicates, we obtain a wonderful variety of filiform and linear growths which may attain to a height of 30 or 40 centimetres. Some are so flexible that the stems bend, falling in curves around the centre of growth, like leaves of grass. If we dilute this same liquid, as it becomes less concentrated the growths are more curved, ramified, dendritic, like those of trees or corals.

In the culture of osmotic growths we may also by appropriate means produce terminal organs resembling flowers and seed-capsules. To do this we wait till the growth is considerably advanced, and then add a large quantity of liquid to the nutrient solution so as to diminish the concentration a hundredfold or more. Spherical {132} terminal organs will then grow out from the ends of the stems, which may during their further growth become conical or piriform in shape.

By superposing layers of liquid of different concentration and decreasing density, one may obtain knots and swellings in the osmotic growths marking the surfaces of separation of the liquid. When a young growth in the vigour of its youth reaches the surface of the water, it spreads out horizontally over the surface of the liquid in thin leaves or foliaceous expansions of different forms.

The preponderating influence in morphogenesis is osmotic pressure, the osmotic forms varying with its intensity, distribution, and mode of application. Whatever the chemical composition of the liquid, similar osmotic forces, modified in the same manner, give rise to forms which have a family resemblance. The chemical nature of the liquid, however, is not entirely without influence on the form. Thus the presence of a nitrate in the mother liquor tends to produce points or thorns. Ammonium chloride in a potassium ferrocyanide solution produces growths shaped like catkins, and the alkaline chlorides tend to produce vermiform growths. {133}

Coralline growths may also be obtained by using appropriate chemical solutions. For this purpose the solution of silicate, carbonate, and dibasic phosphate should be diluted to half strength, with the addition of 2 to 4 per cent. of a concentrated solution of sodium sulphate or potassium nitrate.

Coral-like forms may also be grown from a semi-saturated solution of silicate, carbonate, and dibasic phosphate, to which has been added 4 per cent. of a concentrated solution of sodium sulphate or potassium nitrate. In this we may obtain beautiful growths like madrepores or corals, formed by a central nucleus from which radiate large leaves like the petals of a flower. The presence of nitrate of potassium produces pointed leaves with thorn-like processes recalling the forms of the aloe and the agave.

Most remarkable fungus-like forms may be obtained by commencing the growth in a concentrated solution, and then {134} carefully pouring a layer of distilled water over the surface of the liquid. The resemblance is so perfect that some of our productions have been taken for fungi even by experts. The {135} stem of these osmotic fungi is formed of bundles of fine hollow fibres, while the upper surface of the cap is sometimes smooth, and sometimes covered with small scales. The lower surface of the cap shows traces of radiating lamell? which are sometimes intersected by concentric layers parallel to the outer {136} surface of the cap. In this case the lower surface of the cap shows a number of orifices or canals similar to those seen in many varieties of fungus.

Shell-like osmotic productions may be grown by sowing the mineral in a very shallow layer of concentrated solution, a centimetre or less in depth, and pouring over this a less concentrated layer of solution. By varying the solution or concentration we may thus grow an infinite variety of shell forms. {137}

Capsules or closed shells may be produced in the same way by superimposing a layer of somewhat greater concentration. These capsules consist of two valves joined together at their circumference. The lower valve is thick and strong, while the upper valve may be transparent, translucent, or opaque, but is always thinner and more fragile than the lower one.

Ferrous sulphate sown in a silicate solution gives rise to growths which are green in colour, climbing, or herbaceous, twining in spirals round the larger and more solid calcareous growths.

With salts of manganese, the chloride, citrate or sulphate, the stages of evolution of the growth are distinguished not only by diversities of form, but also by modifications of colour. We may thus obtain terminal organs black or golden yellow in colour on a white stalk. In a similar way we may obtain fungi with a white stalk and a yellow cap, of which the lower surface is black.

Very beautiful growths may be obtained by sowing calcium chloride in a solution of potassium carbonate, with the addition of 2 per cent. of a saturated solution of tribasic potassium phosphate. This will give capsules with figured belts, vertical lines at regular intervals, or transverse stripes composed of projecting dots such as may be seen in many sea-urchins. These capsules are closed at the summit by a cap, forming an operculum, so that they sometimes appear as if formed of two valves. Now and again we may see the upper valve raised by {138} the internal osmotic pressure, showing the gelatinous contents through the opening.

Osmotic productions may be divided into two groups. Some like the silicate growths are fixed. Like vegetables, they develop, become organized, grow, decline, die, and are disintegrated at the spot where they are sown. Others, especially those which are grown in alkaline carbonates and phosphates, have two periods of evolution, the first a fixed period, and the second a wandering {139} one. During the first period their specific gravity is greater than that of the surrounding medium, and they rest immobile at the bottom of the vessel in which they are sown. As they grow, they absorb water and their specific gravity diminishes. Little by little they rise up in the liquid, and finally acquire a considerable amount of mobility, being readily displaced by every current. Hence it is very difficult to photograph these {140} mobile osmotic growths, which swim about in the mother liquor and are often provided with prolongations in the forms of cilia, and sometimes with fins, which undulate as they move. Some of these ciliary hairs are evidently osmotic in their origin, being localized as a tuft at the summit of the growth. Others are apparently crystalline in structure, and are spread over the whole surface of the swimming vesicle. An osmotic growth increases by the absorption of water from a concentrated solution. When the solution is originally saturated it thus becomes supersaturated, and deposits these long ciliary crystals on the surface of the growth.

When a capsule splits in two under the influence of the internal osmotic pressure, it may happen that the operculum or upper valve floats away in the liquid. We thus obtain a free swimming organism, a transparent bell-like form with an undulating fringe, like a Medusa.

Frequently a single seed or stock will give rise to a whole series of osmotic growths. A vesicle is first produced, and then a contraction appears around the vesicle, and this contraction increases till a portion of the vesicle is cut off and swims away free like an amoeba. The same phenomenon may be observed with vermiform growths, a single seed often giving {141} rise in this way to a whole series of amoebiform or vermiform productions.

It must be remembered that in an osmotic growth the active growing portion is the gelatinous contents in the interior, the external visible growth being only a skeleton or shell. We may sometimes succeed in hooking up one of these long vermiform growths, breaking the calcareous sheath, and drawing out a long undulating translucid gelatinous cylinder. The outline of this cylinder is so well defined as to make us doubt whether the fine colloidal membrane which separates it clearly from the liquid can have been formed so rapidly, or if it may not perhaps exist already formed in the interior of its calcareous sheath.

(a) Sodium sulphite.

(b) Potassium bichromate.

(c) Sodium sulphide.

(d) Sodium bisulphite.]

When a large capsular shell such as we have described bursts, it expels a part or the whole of its contents as a gelatinous mass which retains the form of the cavity. Similarly, if we suddenly dilute the mother liquor around an osmotic cell, it bursts by a process of dehiscence, and projects into the liquid a part of its contents, which may thus become an independent vesicle. In this way a single osmotic cell may produce a whole series of independent vesicles.

It is even possible to rejuvenate an osmotic growth that has become degenerate through age. An osmotic production grows old and dies when it has expended the osmotic force contained in the interior of its capsule. A calcium osmotic growth which has thus become exhausted may be rejuvenated by transferring it to a concentrated solution of calcium chloride. It will absorb this, and thus be enabled to renew its evolution and growth when put back again into the original mother liquor. {142}

The structure of osmotic growths is no less varied than their form. Their stems are formed of cells or vesicles juxtaposed, showing cavities separated by osmotic walls. Sometimes the component vesicles have kept their original form, so that the stem has the appearance of a row of beads. Or the cells may be more or less flattened, the divisions being widely separated. Or again, by the absorption of the divisions, a tube may be formed, a veritable vessel or canal in which liquids can circulate. {143}

The foliaceous expansions, or osmotic leaves, also present great varieties both of appearance and of structure. The veins may be longitudinal, fan-shaped, or penniform. We have occasionally met with leaves having a lined or ruled surface, giving most beautiful diffraction colours. The usual structure, however, is vesicular or cellular, as in Fig. 58. In photographs we often get the appearance of lacuna but all these lacuna are closed cavities, the appearance being due to the transparency of the cell walls.

In conclusion we may say that osmotic growths are formed of an ensemble of closed cavities of various forms, containing liquids and separated by osmotic membranes, constituting veritable tissues. This structure offers the closest {144} resemblance to that of living organisms. Is it possible to doubt that the simple conditions which produce an osmotic growth have frequently been realized during the past ages of the earth? What part has osmotic growth played in the evolution of living forms, and what traces of its action may we hope to find to-day? Osmotic growth gives us fibrous silicates, phosphatic nodules, corals, and madrepores; it also gives us formations which remind one of the "atolls," calcareous growths rising like a crown out of the water. The geologist may well consider what role osmotic growth may have played in the formation of the various rocks, siliceous, calcareous, barytic, magnesian, the fibrous and nodular rocks and atolls. The paleontologist relies on the different forms found in his rocks to classify his specimens; from the existence of a shell, he concludes the presence of life. Since, however, forms which are apparently organic may be merely the product of osmotic growth, it is evident that he must reconsider his conclusions. The same may be said of the various forms of coral or of fungoid growths. In the {146} presence of a calcified or silicated fungus we can no longer argue with certainty as to the existence of life, without taking into consideration the possibility that the specimen in question may be an osmotic production.

Whatever our opinion as to its signification, osmotic growth demands the attention of every mind devoted to the study of nature. It is a marvellous spectacle to see a formless fragment of calcium salt grow into a shell, a madrepore, or a fungus, and this as the result of a simple physical force. Why should the study of osmotic growth attract less attention than the formation of crystals, on which so much time and labour has been bestowed in the past?

* * * * *

{147}

CHAPTER XII

THE PHENOMENA OF LIFE AND OSMOTIC PRODUCTIONS--A STUDY IN PHYSIOGENESIS

It is impossible to define life, not only because it is complex, but because it varies in different living beings. The phenomena which constitute the life of a man are far other than those which make up the life of a polyp or a plant; and in the more simple forms life is so greatly reduced that it is often a matter of difficulty to decide whether a given form belongs to the animal, vegetable, or mineral kingdom. Considering the impossibility of defining the exact line of demarcation between animate and inanimate matter, it is astonishing to find so much stress laid on the supposed fundamental difference between vital and non-vital phenomena. There is in fact no sharp division, no precise limit where inanimate nature ends and life begins; the transition is gradual and insensible, for just as a living organism is made of the same substances as the mineral world, so life is a composite of the same physical and chemical phenomena that we find in the rest of nature. All the supposed attributes of life are found also outside living organisms. Life is constituted by the association of physico-chemical phenomena, their harmonious grouping and succession. Harmony is a condition of life.

We are quite unable to separate living beings from the other productions of nature by their composition, since they are formed of the same mineral elements. All the aliments of plants--water, carbon, nitrogen, phosphorus, sulphur--before their absorption and assimilation belonged to the mineral kingdom. The carbon and the water are transformed into {148} sugar and fat, the nitrogen and the sulphur into albumen, and the compounds so formed are then said to belong to the organic world. These organic bodies are returned once again to the mineral world by the action of animals and microbes, which transform the carbon into carbonates, and the nitrogen, sulphur, and phosphorus into nitrates, sulphates, and phosphates. Hence life is but a phase in the animation of mineral matter; all matter may be said to have within itself the essence of life, potential in the mineral, actual in the animal and the vegetable. The flux and reflux of matter is alternate and incessant, from the mineral world to the living, and back again from the living to the mineral world.

At the same time there is a continuous flux of energy. Organic matter contains potential energy, the energy of chemical combination; and during its passage through the living being it is gradually stripped of this energy and returned to the mineral world. The first step in synthetic biology is the addition of potential energy to matter, the reduction of an oxide, the separation of a salt into its radicals, the production of some endothermic chemical combination. The energy stored up by such processes can be again liberated as heat, that fire which the ancients with wonderful prescience long ago recognized as the symbol of life.

Attempts have been made to differentiate a living being by the nature of its chemical combinations, the so-called organic compounds. It was supposed that life alone could realize these and cause the production of the various substances which form the structure of living beings. Of late years, however, a large number of these organic substances have been artificially produced in the laboratory, and the synthetic problems which remain are of the same order as those which have been already solved.

As one learns to know the mineral kingdom and the living world more intimately the differences between them disappear. Thus a living being was supposed to be characterized by its sensibility, i.e. its faculty of reaction against external impressions. But this reaction is a general phenomenon of nature; there is no action without reaction. Neither can the {149} reaction to internal impressions, immediate or deferred, be considered as the characteristic of life, since osmotic growths exhibit a most exquisite sensibility in this direction. Since, then, the faculty of reaction is a general property of matter, the characteristics of life in the lower organisms are only three in number, viz. nutrition, growth, and reproduction by fission or budding. But crystals are also nourished and grow in the water of crystallization. They have moreover a specific form, and every biologist who wishes to establish a parallel between the phenomena of the living and the mineral world is wont to compare living beings with crystals. Crystals, it is said, affect regular geometric forms, salient angles, and rectilinear edges, while living beings have rounded forms without any geometric regularity. Another supposed distinction is that living beings are nourished by intussusception, whereas crystals increase by apposition. Again, living beings are said to assimilate and transform the aliment they absorb, whereas crystals do not transform the matter which is added externally to their structure. Another supposed difference is that living things eliminate and discharge their products of combustion, while the evolution of a crystal is accompanied by no such elimination. Finally, the phenomenon of reproduction is said to be the exclusive characteristic of a living being; but crystals may also be reproduced and multiplied by the introduction of fragments of crystalline matter into a supersaturated solution.

The resemblance between an osmotic growth and a living organism is much closer than that between a living being and a crystal, there being not only an analogy of form, but also of structure and of function. In order to find the physical parallel to life, we must turn to osmosis and osmotic growth rather than to crystals and crystallization.

The first and most striking analogy between living beings and osmotic growths is that of form. The morphogenic power of osmosis gives rise to an infinite variety of forms. An osmotic growth, even at the first sight, suggests the idea of a living thing. One need only glance at the photographs of osmotic productions to recognize the forms of madrepore, fungus, alga, and shell. It is wonderful that a force capable {150} of such marvellous results should have hitherto been almost entirely neglected.

A second analogy between vital and osmotic growths is to be found in their structure, both being formed by groups of cells or vesicles separated by osmotic membranes. An osmotic stem, formed by a row of cellular cavities separated by osmotic membranes, has a great structural resemblance to the knotted stems of bamboos, reeds, and the like. The foliaceous expansions of osmotic growths are formed by colonies of cells or vesicles disposed in regular lines, which may present various patterns of innervation, parallel, palmate, or pennate. Many of the lamellar osmotic growths are striped in parallel lines alternately opaque and transparent. The terminal organs have also their enveloping membranes, their pulp and nucleus, just like vegetable forms.

The analogies of function are no less remarkable than those of form and structure. Nutrition is perhaps the most elementary and essential vital phenomenon, since without nutrition life cannot exist. Nutrition consists in the absorption of alimentary substances from the surrounding medium, the chemical transformation of such substances, their fixation by intussusception in every part of the organism, and the ejection of the products of combustion into the surrounding medium. Osmotic growths absorb material from the medium in which they grow, submit it to chemical metamorphosis, and eject the waste products of the reaction into the surrounding medium. An osmotic growth moreover exercises choice in the selection of the substances which are offered for its consumption, absorbing some greedily and entirely rejecting others. Thus osmotic growths present all the phenomena of nutrition, the fundamental characteristic of life.

In the living organism nutrition results in growth, development, and evolution. Growth and development also follow the absorption and fixation of aliment by an osmotic production. An osmotic production grows, its form develops and becomes more complicated, and its weight increases. An osmotic growth may weigh many hundred times as much as the mineral sown in the solution, the mother liquor losing a {151} corresponding weight. Thus growth, which has hitherto been considered an essential phenomenon of life, is also a phenomenon common to all osmotic productions.

Osmotic growths like living things may be said to have an evolutionary existence, the analogy holding good down to the smallest detail. In their early youth, at the beginning of life, the phenomena of exchange, of growth, and of organization are very intense. As they grow older, these exchanges gradually slow down, and growth is arrested. With age the exchanges still continue, but more slowly, and these then gradually fail and are finally completely arrested. The osmotic growth is dead, and little by little it decays, losing its structure and its form.

The membranes of an osmotic growth thicken with age, and thus oppose to the osmotic exchanges a steadily increasing resistance. Young osmotic cells appear swollen and turgescent, whereas old ones become flaccid, relaxed, and wrinkled. Analogous phenomena are met with in living organisms, the calcareous infiltration of the vessels representing the thickening and hardening of the osmotic membranes. The plumpness of a child and the turgescence of young cells are but the expression of high osmotic tension, while relaxation and flaccidity of the tissues in old age betrays the fall of osmotic pressure in the intracellular tissues.

Circulation of the nutrient fluid may also be observed in an osmotic growth as in a living organism. If we take a calcareous growth with long ramified stems and dilute the mother liquor considerably, we may see currents of liquid issuing from the summit of the growth--currents which are made visible by the cloudy precipitates which they cause. The same current is also rendered visible in the stems themselves by the motion of the granulations and gas bubbles in the interior of the osmotic cells. It is plain that some such circulation must exist, for how could a membrane be formed 30 centimetres from the seed if the membranogenous substance did not circulate through the stem? A moment's consideration will show that the propulsion is due to osmotic pressure and not to mere differences of density, for the liquid {152} which rises in the stem is a concentrated solution of calcium salt much denser than the mother liquor, and the current of liquid after rising in the stem may be seen to fall back again through the liquid.

Organization has long been considered as one of the principal characteristics of life, i.e. the arrangement of matter so as to produce an animated and evolutionary form accompanied by transformation of energy. But osmotic growths are also organizations endowed with the same faculties, and the physical mechanism which is at the basis of their formation is the same as that which determines the organization of living matter.

The phenomena of osmotic growth show how ordinary mineral matter, carbonates, phosphates, silicates, nitrates, and chlorides, may imitate the forms of animated nature without {153} the intervention of any living organism. Ordinary physical forces are quite sufficient to produce forms like those of living beings, closed cavities containing liquids separated by osmotic membranes, with tissues similar to those of the vital organs in form, colour, evolution, and function.

It is only necessary to glance at the photographs of these osmotic growths to appreciate the wonderful variety of form. The variety of function is not less evident, and in many instances, especially with manganese salts, the difference of function of various regions is marked by differences of colour. When a large osmotic cell projects beyond the mother liquor and grows up into the air, it is evident that the function of liquid absorption must be localized in the submerged part. In other cases we have a local evolution of gas, which may be demonstrated by growing a fragment of calcium chloride in a mother liquor composed of the following saturated solutions:--

Potassium carbonate 76 parts. Potassium sulphate 16 " Tribasic potassium phosphate 46 "

During the whole period of growth there is an abundant liberation of bubbles of gas, which is accurately limited to a belt around the base of the growth, and sometimes also to a cap at the summit.

Since morphological differentiations of different parts is but the result of differences of evolution, i.e. of functional differences of the various parts, we may consider that osmotic growths possess the faculty of organization like living beings.

An osmotic growth may be wounded, and a wound delays its growth and development like a disease or an accident in a living being. A wound in an osmotic production may also become cicatrized and covered with a membrane, when the growth will recommence exactly as in a living being.

An osmotic growth is a transformer of energy. It increases in bulk, pushing aside the mother liquor, and thus doing external work. An osmotic growth has a temperature above its medium, since the chemical reaction of which it is the seat is accompanied by the production of heat. We know {154} but little of the transformation of energy which takes place in an osmotic production, but we may say with certainty that it is capable of transforming both chemical energy and osmotic energy into heat and mechanical motion.

An osmotic production is the arena of complicated chemical phenomena which produce a veritable metabolism. It has long been known that diffusion and osmosis may determine various chemical transformations. H. St. Clair Deville has demonstrated that certain unstable salts are partially decomposed by diffusion. Thus during the diffusion of alum, the sulphate of potash is separated from the sulphate of aluminium. Similarly, when the chloride or acetate of aluminium is caused to diffuse, the acids become separated from the aluminia. This decomposition is the result of the different resistance which the medium offers to the diffusion of different ions. This difference of resistance may even cause a difference of potential between two media, similar to the differences of potential in living organisms. Frequently also a difference of hydration in the chemical substances on either side of an osmotic membrane will determine a chemical reaction, which like all other chemical reactions is accompanied by a corresponding transformation of energy. The study of these chemical metamorphoses and the transformations of energy in osmotic growths has opened up a new subject for experimental investigation in the field of organic chemistry.

Coagulation.--There is a most remarkable analogy between the phenomena of coagulation as seen in living beings and the phenomena which occur when the liquid in the interior of an osmotic growth comes into contact with the mother liquor. When the sap of a plant or the blood of an animal escapes into the air or water of the surrounding medium, it coagulates, i.e. it changes from a liquid to a gelatinous consistency. In the same way, when the liquid in the interior of an osmotic growth leaks out into the mother liquor it forms a gelatinous precipitate. This gelatinous precipitation is a physico-chemical phenomenon of the same nature as coagulation. It is by the study of coagulation in liquids less complex than blood that we may hope to elucidate the mechanism of the process, {155} which is simply a physico-chemical phenomenon exactly analogous to gelatinous precipitation. Calcium phosphate is always prone to coagulate; it has been called the gelatinous phosphate of lime, and we have already seen how readily tribasic calcium phosphate takes the form of beautiful transparent colloidal membranes which are gelatinous in texture.

We may obtain colloidal precipitates exactly analogous to coagulated albumin by mixing a weak solution of chloride of calcium with potassium carbonate or tribasic phosphate. Like albumin this precipitate forms flakes, and is deposited slowly as a gelatinous colloidal mass. Like albumin also this calcic solution is coagulated by heat; a solution of a calcic salt of a volatile acid on heating forms a precipitate which has all the appearance of albumin coagulated by heat.

Finally, Arthus and Page have shown that blood does not coagulate when deprived of its calcium salts by the addition of alkaline oxalates, fluorides, or citrates, and that the blood thus treated recovers its coagulability on the addition of a soluble salt of calcium. The coagulation of milk is also a calcium salt precipitation. Coagulation therefore would seem to be merely the colloidal precipitation of a salt of calcium.

Diffusion and osmosis are the elementary phenomena of life. All vital phenomena result from the contact of two colloidal solutions, or of two liquids separated by an osmotic membrane. Hence the study of the physics of diffusion and osmosis is the very basis of synthetic biology.

A living being exhibits two sorts of movements, those which are the result of stimulus from without, and those which are determined by an excitation arising from within. In the higher animals the stimulus or exciting energy coming from the entourage may be infinitely small when compared with the amount of energy transformed. Moreover, the response to an identical excitation may so vary as to give to these different responses an appearance of spontaneity. There is in reality no spontaneity, since the difference in response is governed by previous external impressions which have left their record on the machinery. There is in fact no such thing as a spontaneous action, since every action of a living {156} being has as its ultimate cause a stimulus or excitation coming from without.

The movements of the second category are also conditioned by an excitation, but the stimulus comes from within the organism. These movements consist principally of changes of nutrition, or movements of the circulation and respiration; they are rhythmic in character and are probably produced by the same chemico-physical causes which determine rhythmic movements outside the living body.

Just in the same way osmotic growths present two sorts of movements, external movements and those which are connected with their nutrition. A free osmotic growth swimming in the mother liquor will alter its position and form under the influence of the slightest exterior excitation or vibration. It responds to every variation of temperature, or to a slight difference of concentration produced by adding a single drop of water, and reacts to every exterior influence by displacement or deformation.

An osmotic growth also shows indications of movements which are connected with its nutrition, and these movements are rhythmic, like those of respiration or circulation in a living organism. The growth of an osmotic production shows itself not as a continuous process but periodically. The water traverses the membrane, raises the pressure, and distends the cell; at first the cell wall resists by reason of its elasticity, it then suddenly relaxes, yielding to the osmotic pressure and bulging out at a thinner spot on the surface; the internal pressure falls suddenly, and there is a pause in the growth.

This rhythmic growth may be best observed by sowing in a solution of a tribasic alkaline phosphate, pellets composed of powdered calcium chloride moistened with glycerine, to which has been added 1 per cent. of monobasic calcium phosphate. The experiment is so arranged as to bend or incline the growing stems which shoot out from these grains. This may be done by carefully pouring above the mother liquor a layer of water, or a less concentrated solution. As the internal osmotic pressure rises, the drooping extremity of the twig will become turgescent and gradually lift itself {157} up, and then suddenly fall again for several millimetres. We have frequently watched this rhythmic movement for an hour or more--a slow gradual elevation of the extremity of the twig and a rapid fall recurring every four seconds or so.

It may be objected that the substance of an osmotic growth is continually undergoing change, whereas a living organism transforms into its own substance the extraneous matter which it borrows from its environment. The distinction, however, is only an apparent one. The substance of a living being is also continually undergoing chemical change; it does not remain the same for a single instant. We see an evidence of this change in the evolution of age; the substance of the adult is not that of the infant. In some living organisms such as insects, especially the ephemerid?who have but a brief existence, this change of substance is even more rapid than that in an osmotic growth.

It has been objected that osmotic productions cannot be compared with living organisms since they contain no albuminoid matter. This is to consider life as a substance, and to confound the synthesis of life with that of albumin. If albumin is ever produced by synthesis in the laboratory it will probably be dead albumin. All living organisms contain albumin; this is probably due to the fact that albuminoid matter is particularly adapted for the formation of osmotic membranes. Our osmotic productions are composed of the same elements as those which constitute living beings; an osmotic growth obtained by sowing calcium nitrate in a solution of potassium carbonate with sodium phosphate and sulphate contains all the principal elements of a living organism, viz. carbon, oxygen, hydrogen, nitrogen, sulphur, and phosphorus.

The whole of the vegetable world is produced by the osmotic growth of mineral substances, if we except the small amount of organic matter contained in the seeds.

The most important problem of synthetic biology is not so much the synthesis of the albuminoids as the reduction of carbonic acid. In nature this reduction is accomplished by the radiant energy of the sun, by the agency of the catalytic action of chlorophyll. {158}

The physico-chemical study of osmotic growth is as yet hardly begun; we have but indicated the method, the way is open, and the problems awaiting solution are legion. Only work and ever more work and workers are required. Experiments should be made with substances which are chemically unstable like the albuminoids, substances which readily combine and dissociate again, alternately absorbing and giving up the potential energy which is the essence of life. Experiments should also be made with substances which readily unite or decompose under the influence of water, since hydration and hydrolysis appear to be the dominant mechanism in all vital reaction, as they undoubtedly are in osmotic growth, which consists of an increase of hydration on one side of an osmotic membrane and a diminution on the other side.

Life is not a substance but a mechanical phenomenon; it is a dynamic and kinetic transference of energy determined by physico-chemical reactions; and the whole trend of modern research leads to the belief that these reactions are of the same nature as those met with in the organic world. It is the grouping of physical reactions and their mode of association and succession, their harmony in fact, which constitutes life. The problem we have to solve in the synthesis of life is the proper attuning and harmonizing of these physical phenomena, as they exist in living beings, and there should be no absolute impossibility in our some day realizing this harmony in whole or in part.

Albert Gaudry says: "I cannot conceive why in determining the connecting links of the animal world the fact that an organic body is formed of such and such elements should be of greater importance than the manner in which these elements are grouped. Descartes regarded extension as the essential property of an organized being; he supposed it to be inert of itself, and that it had the Deity for its motive force. To-day the hypothesis of Descartes has given way to that of Leibnitz, who regards force as the essential property of the living being, the visible and tangible matter being only of secondary importance. If we regard the living being as a force, this force is able to aggregate matter under such and such a form, {159} with such or such a structure, and such or such a chemical essence. It does not seem that the classification depending on differences of substance are any more important than those which depend on differences of form."

The biological interest of osmotic productions is quite independent of the chemical nature of the substances which enter into their growth. All substances which produce osmotic membranes by the contact of their solutions exhibit phenomena analogous to those of nutrition. Osmotic morphogenesis is a physical phenomenon resulting from the contact of the most diverse substances. It has given us our first glimpse of the manner in which a living being may be supposed to have been formed according to the ordinary physical laws of nature. We cannot at present produce osmotic growths with all the combinations found in living beings, but that is only because chemistry still lags far behind physics in the synthesis of organic forms.

We are often told "not to force the analogy." But error is equally produced by the exaggeration of unimportant differences. We have already seen that nutrition, absorption, transformation, and excitation are not the characteristics of living organisms alone; nor is reaction to external impressions the appanage only of animate beings. To insist on the resemblance between an osmotic production and a living being is not to force an analogy but to demonstrate a fact.

Let us briefly recapitulate. An osmotic growth has an evolutionary existence; it is nourished by osmosis and intussusception; it exercises a selective choice on the substances offered to it; it changes the chemical constitution of its nutriment before assimilating it. Like a living thing it ejects into its environment the waste products of its function. Moreover, it grows and develops structures like those of living organisms, and it is sensitive to many exterior changes, which influence its form and development. But these very phenomena--nutrition, assimilation, sensibility, growth, and organization--are generally asserted to be the sole characteristics of life.

* * * * *

{160}

CHAPTER XIII

EVOLUTION AND SPONTANEOUS GENERATION

By many biologists, even at the present day, the origin and evolution of living beings is considered to be outside the domain of natural phenomena, and hence beyond the reach of experimental research. The change in our views on this subject is due to a Frenchman, Jean Lamarck, who was the true originator of the scientific doctrine of evolution. At a time when the miraculous origin of every living being was regarded as an unchangeable verity, and was defended like a sacred dogma, Lamarck boldly formulated his theory of evolution, with all its attendent consequences, from spontaneous generation to the genealogy of man.

In his Philosophie Zoologique, which appeared in 1809, Lamarck put forth his claim to regard all the phenomena of life, of living beings, and of man himself as pertaining to the domain of natural phenomena. According to him, all bodies which are met with in nature, organic and inorganic alike, are subject to the same laws. Life is a physical phenomenon, and all the processes of life are due to mechanical causes, either physical or chemical. He writes: "?leur source le physique et le moral ne sont sans doute qu'une seule et mete chose. Il faut rechercher dans la consideration de l'organisation les causes metes de la vie."

In the intellectual evolution of the human mind perhaps no advance has been more important than that of Lamarck--the conquest of the domain of life by human intelligence. In conformity with the true scientific method, he founds his doctrine on the facts and phenomena of nature. "I confine myself," he says, "within the bounds of a simple contemplation {161} of nature." It was this observation of the gradual perfecting of living organisms from the simplest to the most complicated that inspired Lamarck with the idea of evolution and transformation. "How," he says, "can we help searching for the cause of such wonderful results? Are we not compelled to admit that nature has produced successively bodies endowed with life, proceeding from the simplest to the most complex?"

The various products of nature have been divided into classes, genera, and species, simply to facilitate their study. Modern research tends to show that there is no definite line of demarcation even between the animal, vegetable, and mineral kingdoms. All our classification is artificial, and the passage from one division to another is gradual and insensible. Lamarck expresses this idea very clearly: "We must remember that classes, orders, and families, and all such nomenclature, are methods of our own invention. In nature there are no such things as classes or orders or families, but only individuals. As we become better acquainted with the productions of nature, and as the number of specimens in our collections increases, we see the intervals between the classes gradually fill up, and the lines of separation become effaced."

Lamarck also raises his voice against the supposed immutability of species. "Species have only a relative constancy, depending on the circumstances of the individuals. The individuals of a given species perpetuate themselves without variation only so long as there is no variation in the circumstances which influence their existence. Numberless facts prove that when an individual of a given species changes its locality, it is subjected to a number of influences which little by little alter, not only the consistency and proportions of its parts, but also its form, its faculty, and even its organization; so that in time every part will participate in the mutations which it has undergone."

Lamarck also clearly affirms the fact of spontaneous generation. "I hope to prove," he says, "that nature possesses means and faculties for the production of all the forms which we so much admire. Rudimentary animals and plants have {162} been formed, and are still being formed to-day, by spontaneous generation."

Lamarck himself gives a summary of his doctrine in the following six propositions:--

1. "All the organized bodies of our globe are veritable productions of Nature, which she has successively formed during the lapse of ages.

2. "Nature began, and still recommences day by day, with the production of the simplest organic forms. These so-called spontaneous generations are her direct work, the first sketches as it were of organization.

3. "The first sketches of an animal or a vegetable growth being begun under favourable conditions, the faculties of commencing life and of organic movement thus established have gradually developed little by little the various parts and organs, which in process of time have become diversified.

4. "The faculty of growth is inherent in every part of an organized body; it is the primary effect of life. This faculty of growth has given rise to the various modes of multiplication and regeneration of the individual, and by its means any progress which may have been acquired in the composition and forms of the organism has been preserved.

5. "All living things which exist at the present day have been successively formed by this means, aided by a long lapse of time, by favourable conditions, and by the changes on the surface of the globe--in a word, by the power which new situations and new habits have of modifying the organs of a body which is endowed with life.

6. "Since all living things have undergone more or less change in their organization, the species which have been thus insensibly and successively

produced can have but a relative constancy, and can be of no very great antiquity."

The admirable work of Lamarck was absolutely neglected in France, where it was treated as unworthy even of consideration. This neglect profoundly afflicted Lamarck, who gradually sank a victim to the opposition of his contemporaries. He left, however, one disciple, Etienne Jeoffroy St. {163} Hilaire, but he too was soon reduced to silence under the weight of authority of his adversaries.

Before the doctrine of evolution could live and take its proper place, it had to be reborn in England--the country of liberty. This resuscitation was due to Darwin, who added to it his illuminating doctrine of natural selection. But apart from this and a perfecting of its various details, Lamarck had already formulated the doctrine of evolution with perfect precision. Lamarck's work was still-born, whereas that of Darwin lived and grew to its full development. This was due, not to any imperfection or insufficiency in Lamarck's work, but {164} to the milieu into which it was born. It was the environment that stifled the offspring of Lamarck.

In 1868, Ernest Haeckel speaks of the genius of Lamarck in these words: "The chief of the natural philosophers of France is Jean Lamarck, who takes his place beside Goethe and Darwin in the history of evolution. To him belongs the imperishable glory of being the first to formulate the theory of descent, and of founding the philosophy of nature on the solid basis of biology," and adds, "There is no country in Europe where Darwin's doctrine has had so little influence as in France." Haeckel has but done tardy justice in his discovery of and testimony to the genius of Lamarck.

The spirit of opposition does not seem to have much changed in France since Lamarck's time. In 1907 the Academie des Sciences de Paris excluded from its Comptes Rendus the report of my researches on diffusion and osmosis, because it raised the question of spontaneous generation.

The majority of scientists seem to consider that the question of spontaneous generation was definitely settled once for all when Pasteur's experiments showed that a sterilized liquid, kept in a closed tube, remained sterile.

Without the idea of spontaneous generation and a physical theory of life, the doctrine of evolution is a mutilated hypothesis without unity or cohesion. On this point Lamarck speaks most clearly: "Although it is customary when one speaks of the members of the animal or vegetable kingdom to call them products of nature, it appears that no definite conception is attached to the expression. Our preconceived notions hinder us from recognising the fact that Nature herself possesses all the faculties and all the means of producing living beings in any variety. She is able to vary, very slowly but without cessation, all the different races and all the different forms of life, and to maintain the general order which we see in all her works."

The doctrine of Lamarck is frequently misinterpreted. We often hear it expressed as "Function makes the organ," or even "Function creates the organ." This is equivalent to saying, "Life makes the living being," which is incomprehensible, {165} making of function a sort of immaterial and independent entity which constructs a material organ in order to lodge within it. No such idea is to be found in all the works of Lamarck. He formulates his law in the following terms: "In every animal which is still undergoing development, the frequent and sustained use of any one organ increases its size and power, whereas the constant neglect of the use of such organ weakens and deteriorates it, so that it finally disappears."

In his expression of this law Lamarck insists on the fact that organization precedes function. He affirms only that function, i.e. action and reaction, modifies the organ; or, in other words, that organisms are modelled by the action of exterior forces acting upon them. It is in this sense only that function may be said to make an organ, but this mode of expression should be avoided, as it is apt to be misunderstood.

Astronomy teaches us that our globe was detached from the sun in an incandescent state, and geology asserts that this earth has passed through a period of long ages when its temperature was incompatible with the existence of life. It was only with the cooling of the earth crust that it was possible for living beings to make their appearance. Hence they must of necessity have been produced spontaneously from terrestrial material under the influences of chemical and physical forces. This opinion imposes itself on all who reflect and judge freely. In the same way the doctrine of evolution necessitates as a corollary the doctrine of spontaneous generation. The doctrine of evolution should reconstitute every link in the chain of beings from the simplest to the most complicated; it cannot afford to leave out the most important of all, viz. the missing link between the inorganic and the organic kingdoms. If there is a chain, it must be continuous in all its parts, there can be no solution of continuity.

Evolutionists like Lamarck and Haeckel admit spontaneous generation, not as the most probable, but as the only possible explanation of the phenomenon of life.

Lamarck shows us the apparition of living things at a certain epoch of the earth's evolution, and the gradual {166} development of more complicated forms as the conditions changed on the surface of the globe. Darwin shows how heredity and natural selection tend to accentuate the variations which are favourable to existence. Haeckel demonstrates the parallelism between ontogenesis and philogenesis--between the successive forms in the evolution of the embryo and the successive forms of the individual in the evolution of a race. These are great and admirable conquests of the human intelligence, they have demonstrated the first appearance and the progressive evolution of living beings; it now only remains for us to explain them.

The doctrine of evolution, while enforcing the fact of spontaneous generation and progressive evolution, gives us no hint as to the physical mechanism of such generation. It does not tell us by what forces, or according to what laws, the simpler forms of life have been produced, or in what manner differences of environment have acted in order to modify them. The doctrine asserts the simultaneous variations in organic forms and in the physical influences which produce them, but says {167} nothing as to their mode of action. The Darwinian theory shows how acquired variations are transmitted and accentuated by natural selection, but it says nothing as to how these variations may be acquired. In the same way we are in entire ignorance as to the physical mechanism of ontogenetic development, the evolution of the embryo.

The morphogenic action of diffusion produces osmotic growths of extreme variety. Most of these forms recall those of living things--shells, fungi, corals, and algae. The analogy of function is quite as close as the resemblance of form. The study of osmosis, however, is as yet in its infancy, and osmotic productions vary with the physical conditions of chemical constitution, temperature, concentration, and the like. The study of the organizing action of osmosis on organic material has as yet been hardly attempted.

Osmosis produces growths of great complexity, much more complicated indeed than the more simple forms of living organisms. This marvellous complexity of an osmotic growth may be compared with another fact, the ontogenetic development of the ovum, a single cell which under favourable conditions of environment may evolve into a most complicated organism. These considerations lead to the belief that the beginning of life has not been the production of a simple primitive form from which all others are descended, but that a number of such primitive forms may have been produced, forms which by a rapid physical development attained a high degree of complexity. Osmotic morphogenesis shows us that the ordinary physical forces have in fact a power of organization infinitely greater than has been hitherto supposed by the boldest imagination.

When we consider the ignorance in which we still remain as to the phenomena which pass before our very eyes, how can we expect to understand those which occurred in past ages, when the physical and chemical conditions were so immensely different from those which obtain in our own time? What do we know even now of the physical and chemical phenomena which take place in the unfathomed depths of the ocean, where for aught we know even at the present time the same {168} process may be going on--the genesis of life, and the emergence of living beings out of the inanimate mineral world? "Even now," says Albert Gaudry, "polyps and oceanic animalcul?are building up vast coral reefs and rocks. The oxygen and hydrogen which existed once was water, the oxygen and nitrogen which once made air, the carbon, the phosphorus, the silica and the lime which once were solid rock, now form the substance of living beings. The silica is deposited in the skeleton of a sponge or a radiolaria, the shell of a foraminifera or the carapace of a crustacean, or unites with phosphorus to form the bones of a vertebrate. A very tumult of life has succeeded to the primitive silence of inert matter. Life has invaded the earth, and we see on all sides the inanimate mineral kingdom being changed into a living world."

The admission that life may have appeared on the earth under the influence of natural forces and according to physical laws and conditions different from those of the present era throws a vivid light on the study of biogenesis, spontaneous generation, and evolution. The means of research are now indicated, and we have only to study the documents already in our possession in order to know the conditions which obtained when life first appeared on the globe. We must endeavour to reproduce these conditions and to study their effects.

Since all living beings are formed of the same elements as those of the mineral world, the term "organic" as applied to combinations can only be used in order to emphasize the complexity of their constitution. It was formerly believed that these organic combinations were the result of life, and could not be reproduced except by living organisms. To-day many of these organic substances are produced in the laboratory from inorganic materials. In the past history of the globe it is easy to imagine conditions which would facilitate the synthesis of organic substances without the interposition of life. At the temperature of the electric furnace, which was that of the earth at an early period of its evolution, chemical combinations are possible quite other than those obtaining under the present conditions of temperature and pressure. At the higher temperature of the early {169} geological era, silicides, carbides, phosphides, and nitrides were formed in stable combinations instead of the oxides, silicates, carbonates, phosphates, and nitrates of the present time. These combinations existed on the earth at a time when the conditions of temperature precluded the existence of water in a liquid state. As the temperature cooled, and the water vapour became condensed, it entered into chemical combination with the various rocks, producing organic compounds like acetylene, which results from the action of water on calcium carbide. H. L 闓 icque has developed a theory as to the formation of various rocks under these conditions, which he communicated in 1903 to the French Society of Civil Engineers.

The chemical evolution of the globe has undergone great changes as the temperature gradually fell and the constitution of its crust altered. As long as the temperature was higher than that at which water can exist, all chemical reactions must have taken place between anhydric substances, elements and salts in a state of fusion. These conditions are very different from those of the present-day chemistry, which is the chemistry of aqueous solutions. We may hope to be able to reproduce the earlier conditions by the experimental study of anhydric substances in a state of fusion.

At a later period, that of the primary and secondary rocks, there was a uniform and constant temperature of about 40?C. The atmosphere was charged with water vapour, and all the conditions were present for the production of storms and tempests. The atmosphere during long ages must have been the seat of formidable and incessant electric discharges; these discharges are the most powerful of all physical agents of chemical synthesis, and will cause nitrogen to combine directly to form various compounds-- nitrates, cyanides, and ammonia. Carbonic acid would also be present in abundance and would enter into combination with these nitrogenous compounds. In this way we may imagine that compounds were formed which by some process of physical synthesis subsequently gave rise to vast quantities of albuminoid matter. At that time the seas and oceans contained all those substances which have {170} since been fixed by the metamorphism of the primitive rocks, or deposited in the sedimentary strata. Most of the elements in our minerals were formerly in a state of solution in these primeval seas, which contained carbonates, silicates, and soluble phosphates in great abundance. As the crust gradually cooled, the terrestrial atmosphere of necessity altered in composition, and the slow evolution of the atmosphere no doubt also exercised an influence on the development of living beings.

Paleontology teaches us that the earliest living organism appeared in the sea. The most ancient of living things, those of the primary ages, which were of greater duration than all other ages put together, were all aquatic. We find moreover that every living organism consists of liquids, solutions of crystalloids and colloids separated by osmotic membranes; and it is significant that the ocean, that vast laboratory of life, is also a solution of crystalloids and colloids. It is evident, then, that we must look to the study of solutions if we would hope to discover the nature and origin of life.

Life is an ensemble of functions and of energy-transformations, an ensemble which is conditioned by the form, the structure, and the composition of the living being. Life, therefore, may be said to be conditioned by form, i.e. the external, internal, and molecular forms of the living being.

All living things consist of closed cavities, which are limited by osmotic membranes, and filled with solutions of crystalloids and colloids. The study of synthetic biology is therefore the study of the physical forces and conditions which can produce cavities surrounded by osmotic membranes, which can associate and group such cavities, and differentiate and specialize their functions. Such forces are precisely those which produce osmotic growths, having the forms and exhibiting many of the functions of living beings. Of all the theories as to the origin of life, that which attributes it to osmosis and looks on the earliest living beings as products of osmotic growths is the most probable and the most satisfying to the reason.

We have already seen that the seas of the primary and {171} secondary ages presented in a high degree the particular conditions favourable for the production of osmotic growths. During these long ages an exuberant growth of osmotic vegetation must have been produced in these primeval seas. All the substances which were capable of producing osmotic membranes by mutual contact sprang into growth,--the soluble salts of calcium, carbonates, phosphates, silicates, albuminoid matter, became organized as osmotic productions,--were born, developed, evolved, dissociated, and died. Millions of ephemeral forms must have succeeded one another in the natural evolution of that age, when the living world was represented by matter thus organized by osmosis.

The experimental study of osmotic morphogeny adds its weight of evidence in the same direction. When we see under our own eyes the cells of calcium become organized, develop and grow in close imitation of the forms of life, we cannot doubt that such a transformation has often occurred in the past history of our planet, and the conviction becomes irresistible {172} that osmosis has played a predominant role in the history of our earth and its inhabitants. It is a matter of astonishment that the scientist has taken no notice of the active part which osmosis has played in the evolution of our earth. On the effects of this most important physical phenomenon science has hitherto remained entirely mute.

* * * * *